Algorithmic Graph Theory

James A. McHugh

New Jersey Institute of Technology

PRENTICE HALL, *Englewood Cliffs, New Jersey 07632*

Library of Congress Cataloging-in-Publication Data

McHugh, James A.
    Algorithmic graph theory / by James A. McHugh.
        p.   cm.
    Bibliography: p.
    Includes index.
    ISBN 0-13-023615-2
    1. Graph theory. 2. Graph theory--Data processing. I. Title.
QA166.M39 1990
511'.5--dc19   88-30722
                                                       CIP

Editorial/production supervision and
    interior design: Joe Scordato
Cover design: Lundgren Graphics, Ltd.
Manufacturing buyers: Mary Noonan and Bob Anderson

Dedicated to my parents: Ann and Peter

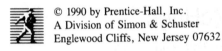

© 1990 by Prentice-Hall, Inc.
A Division of Simon & Schuster
Englewood Cliffs, New Jersey 07632

All rights reserved. No part of this book may be
reproduced, in any form or by any means,
without permission in writing from the publisher.

Printed in the United States of America

10 9 8 7 6 5 4 3 2 1

ISBN 0-13-023615-2

PRENTICE-HALL INTERNATIONAL (UK) LIMITED, London
PRENTICE-HALL OF AUSTRALIA PTY. LIMITED, Sydney
PRENTICE-HALL CANADA INC., Toronto
PRENTICE-HALL HISPANOAMERICANA, S.A., Mexico
PRENTICE-HALL OF INDIA PRIVATE LIMITED, New Delhi
PRENTICE-HALL OF JAPAN, INC., Tokyo
SIMON & SCHUSTER ASIA PTE. LTD., Singapore
EDITORA PRENTICE-HALL DO BRASIL, LTDA., Rio de Janeiro

Preface

This book covers graph algorithms, pure graph theory, and applications of graph theory to computer systems. The algorithms are presented in a clear algorithmic style, often with considerable attention to data representation, although no extensive background in either data structures or programming is needed. In addition to the classical graph algorithms, many new random and parallel graph algorithms are included. Algorithm design methods, such as divide and conquer, and search tree techniques are emphasized. There is an extensive bibliography, and many exercises. The book is appropriate as a text for both undergraduate and graduate students in engineering, mathematics or computer science, and should be of general interest to professionals in these fields as well.

Chapter 1 introduces the elements of graph theory and algorithmic graph theory. It covers the representations of graphs, basic topics like planarity, matching, hamiltonicity, regular and eulerian graphs, from theoretical, algorithmic, and practical perspectives. Chapter 2 overviews different algorithmic design techniques, such as backtracking, recursion, randomization, greedy and geometric methods, and approximation, and illustrates their application to various graph problems.

Chapter 3 covers the classical shortest-path algorithms, an algorithm for shortest paths on euclidean graphs, and the Fibonacci heap implementation of Dijkstra's algorithm. Chapter 4 presents the basic results on trees and acyclic digraphs, a minimum spanning tree algorithm based on Fibonacci heaps, and includes many applications, such as register allocation, deadlock avoidance, and merge and search trees.

Chapter 5 gives an especially thorough introduction to depth-first search and the classical graph structure algorithms based on depth first search, such as block and strong component detection. Chapter 6 introduces both the theory of connectivity and network flows and shows how connectivity and diverse routing problems may be solved using flow techniques. Applications to reliable routing in unreliable networks and to multiprocessor scheduling are given.

Chapters 7 and 8 introduce coloring, matching, vertex and edge covers, and allied concepts. Applications to secure (zero-knowledge) communication, the design of deadlock free systems, and optimal parallel algorithms are given. The Edmonds matching algorithm, introduced for bipartite graphs in Chapter 1, is presented here in its general form.

Chapter 9 presents a variety of parallel algorithms on different architectures, such as systolic arrays, tree processors, and hypercubes, as well as for the shared memory model of computation. Chapter 10 presents the elements of complexity theory, including an introduction to the complexity of random and parallel algorithms.

I greatly appreciate the help given to me in the preparation of this book by a number of graduate students who read and helped correct earlier versions of the manuscript. These include: Krishna Ayala, Jiann-Ru Chiou, Yaw-Nan Duh, Michael Halper, Shun-Hsien Huang, Pankaj Kumar, and Lai-Wu Luo. A State of New Jersey SBR grant provided support during the initial period of the work. I would also like to thank Jim Fegen and Joe Scordato of Prentice Hall for their help in the development and preparation of this book. Finally, I appreciate the last careful reading of the manuscript by Peter and Jimmy, and the constant support I received from my wife, Alice.

Preface

Contents

Figure 3-18a illustrates the effect of a subsequent deletemin on this nine-element heap. Figure 3-18b illustrates the effect of a decrease-key operation, where the key of the item with key equal to 8 is reduced to 4.5. Figure 3-16c illustrates the effect of reducing the key 7 to 2.5. Observe that the item with key 6 has lost two children by this point, which triggers a cascade-cut of the key 6 item from its parent, the tree root with key 2. A final deletemin is shown in Figure 3-18d, at which point there are five nodes on the root list.

REFERENCES AND FURTHER READING

See Clarkson (1987) for the visibility graph, where faster algorithms are described. The euclidean shortest path algorithm is from Sedgewick and Vitter (1986) who also give experimental results. See Fredman and Tarjan (1987) for Fibonacci heaps and their application to Dijkstra's algorithm. See Tarjan (1983) for a sophisticated discussion of shortest path problems and further references. Skiscim and Golden (1987) present generalized Dijkstra's algorithms for the k shortest paths in a euclidean network. Lipton and Tarjan (1977) give a planar shortest path algorithm; see Melhorn (1984) for a discussion. See Edmonds and Fulkerson (1968) and Edmonds and Karp (1972) for the minimax path edge length generalization of Dijkstra's problem and Aho, Garey, and Ullman (1972) for the transitive reduction problem, both of which are referred to in the exercises. Italiano (1986) gives an algorithm for transitive closure in a dynamic environment using lazy insertion that has amortized performance $O(n \log n)$ for a digraph of order n.

EXERCISES

1. Modify Dijkstra's algorithm to find all shortest paths between a given pair of vertices.
2. Construct examples where Dijkstra's algorithm works correctly in the presence of a negative edge(s) and where it works incorrectly in the presence of a negative edge(s).
3. How could you adapt Dijkstra's algorithm to the case where the vertices as well as the edges have weights which contribute to the "length" of a path?
4. Let $G(V, E)$ be a weighted digraph as in Dijkstra's problem, and let x and y be the source and target vertices for the problem. If a new edge is added to G, can the information provided by the Dijkstra traversal be efficiently updated, that is, without running the whole algorithm again? Consider as well the related problems of accounting for a change in the weight of an existing edge, or the deletion of an edge.
5. Consider the same problem as in the previous question, except that now the modification is that a new vertex (and its set of incident edges) is added to G. Consider as well the related problem of deleting an existing vertex.
6. There are N^2 shortest paths in a weighted digraph of order N. If the weight of a single edge changes, what is the maximum number of shortest paths that are affected?
7. Modify Dijkstra's algorithm to find a shortest path between x and y such that the length of the maximum length edge on the path is minimized (Edmonds and Fulkerson (1968) and Edmonds and Karp (1972)).
8. How well does the following greedy shortest path algorithm perform: always extend the estimated shortest-path-tree by adding to it the shortest edge from the tree to a nontree vertex.
9. Design an algorithm for finding the shortest cycle in a positive weighted digraph.

10. What difference(s) are there in the way Floyd's and Ford's algorithms react to negative cycles?

11. If the shortest path from i to j is not unique, which path is selected by Floyd's algorithm?

12. What graphical objects or invariants do we calculate if we replace min by max in Floyd's algorithm?

13. Adapt Floyd's algorithm to find shortest paths avoiding a given set of vertices.

14. Can you adapt Floyd's algorithm to find a shortest cycle on a given vertex? Explain.

15. Suppose $G(V, E)$ has a cut-vertex. How can you adapt Floyd's algorithm to capitalize on the structure of the graph?

16. Prove the following specialized version of Floyd's algorithm (Warshall's algorithm) correctly computes the transitive closure of a digraph $G(V, E)$:

```
for k  =  1 to |V| do
   for i  =  1 to |V| do
      for j  =  1 to |V| do
         Set A(i,j) to ( A(i,j) or (A(i,k) and A(k,j)) ),
```

where **and/or** bit operations are performed on the $|V|$ by $|V|$ adjacency matrix **A** interpreted as a bit matrix.

17. What happens if Ford's algorithm is applied to a graph? What if the edge weights are restricted to be positive?

18. The *transitive closure* of $G(V, E)$ has an edge between every pair of vertices that are connected in G. The *transitive reduction* of $G(V, E)$ is a digraph with the same transitive closure as G, but with as few edges as possible. We can also define a *minimal reduction* of G as a digraph with the same transitive closure as G and containing no smaller digraph with the same property. Is a transitive reduction unique? Is a minimal reduction unique? What is the relation between a transitive reduction and a minimal reduction? Can you design an algorithm to find a transitive reduction (Aho, Garey, and Ullman (1972))?

19. Suppose each edge weight in a weighted graph gives the probability of failure of the edge and that we wish to find the most reliable path between a given pair of vertices. That is, find a path with the least probability of failure. Could Dijkstra's or the other shortest path algorithms be adapted to this problem?

20. Compare the performance of Dijkstra's algorithm with that of the Sedgewick-Vitter algorithm for a class of randomly generated connected euclidean graphs.

21. Design an algorithm to find the k shortest paths in a network with positive edge weights.

4

Trees and Acyclic Digraphs

4-1 BASIC CONCEPTS

A *tree* is an acyclic connected graph. The following theorem summarizes the basic properties of trees:

Theorem (Characterizations of Trees). Let $G(V, E)$ be a graph. Then G is a Tree if and only if one of the following properties holds:

(1) G is connected and $|E| = |V| - 1$,
(2) G is acyclic and $|E| = |V| - 1$,
(3) There exists a unique path between every pair of vertices in G.

The proof of this theorem is straightforward.

An *endpoint* of a tree is a tree vertex of degree one. A nontrivial tree has from 2 to $|V| - 1$ endpoints. A *center* of a tree is a tree vertex of minimum eccentricity. A tree has exactly 1 or 2 centers. A *star* is a tree of diameter 1 or 2. An arbitrary acyclic graph is called a *forest*. A forest with k components has $|V| - k$ edges.

A *rooted tree* is a tree in which we identify a distinguished vertex v which we call the root. We then define the *level* of a vertex: vertices at distance i from the root lie at Level $i + 1$; v itself lies at Level 1. The *height* of the rooted tree is defined as its maximum level.

Directed trees. We call a digraph a *tree* if its underlying undirected graph is a tree in the undirected sense. We call a vertex v a *root* of a digraph G if there are directed paths from v to every other vertex in G. We call a digraph a *directed tree* or *arborescence* if it is a tree and contains a root. We call a vertex of out-degree zero in a directed tree an *endpoint*.

We use genealogical terminology to describe the relations between the vertices in a directed tree. Thus, if (u, v) is a (directed) edge, then u is called the *parent* or *father* of v, and v is called the *child* or *son* of u. If there is also an edge (u, w), then v and w are called *siblings*. If there is a path from a vertex u to a vertex v, then u is called an *ancestor* of v, and v is called a *descendant* of u. The subgraph of a directed tree T induced by a vertex u and all its descendants is called the *subtree rooted at u* and is denoted by $T(u)$.

Ordered trees. An *ordered tree* is a directed tree in which the set of children of each vertex is ordered. A *binary tree* is an ordered tree in which no vertex has more than two children. One of the children is called the *left child*, while the other is called a *right child*. The subtree rooted at the left child of v is called the *left subtree* of v, and the one at the right is the *right subtree* of v. In a *complete binary tree*, every vertex has either two children or none. In a *balanced complete binary tree*, every endpoint has the same level. We have:

Theorem (Complete Binary Trees). A complete balanced binary tree of height h has $2^h - 1$ vertices. Equivalently, a complete balanced binary tree of order $|V|$ (where $|V| + 1$ is a power of 2) has height $\lg(|V| + 1)$.

An *N-ary tree* is a generalization of binary trees where we allow each vertex to have as many as N ordered children. In a *complete balanced N-ary tree*, every endpoint has the same level. We have

Theorem (Complete N-ary Trees). A complete balanced N-ary tree of height h has

$$(N^h - 1)/(N - 1) \text{ vertices}.$$

Equivalently, a complete balanced N-ary tree of order $|V|$ (where $(N - 1)|V| + 1$ is a power of N) has height

$$\log_N((N - 1)|V| + 1).$$

Spanning trees, fundamental cycles, and minimal cuts. There are simple, but important and well-known relations between the spanning trees, cycles, and edge disconnecting sets of a graph. To describe these relations, we will introduce some terminology. Let $G(V, E)$ be a connected graph and let T be a spanning tree of G. An edge of G not lying in T is called a *chord* of T. Each chord of T determines a cycle in G, called a *fundamental cycle;* namely, the cycle caused by adding the chord to T. Any cycle in G can be represented in terms of these fundamental cycles. Let us define the *symmetric difference* of two paths P_1 and P_2 (including closed paths) as the graph that results by removing any edges that P_1 and P_2 have in common as well as any isolated vertices that result after the removal of these edges.

Theorem (Fundamental Cycle Representation). Let $G(V, E)$ be a connected graph and let T be a spanning tree of G. Then, every cycle c in G can be represented as the symmetric difference of the fundamental cycles determined by the edges of c which are chords of T.

There is an analogous theorem for certain sets of edges that disconnect G. Let $G(V, E)$ be a connected graph. An *edge disconnecting set* is a set of edges whose removal disconnects G. A *minimal edge disconnecting set* is an edge disconnecting set not properly containing another edge disconnecting set. A minimal edge disconnecting set is also called a *cutset*. Equivalently, a cutset is an edge disconnecting set whose removal disconnects G into exactly two components. If X is a nonempty set of vertices of

G, the set of edges in $E(G)$ of the form $\{(x, x') \,|\, x \text{ in } X, x' \text{ in } X^c\}$ is called a *cut* and is denoted by (X, X^c). A cut which does not properly contain another cut is called a *minimal cut*. Equivalently, a minimal cut is a cut whose removal disconnects G into exactly two components. Minimal cuts are cutsets, and conversely.

Just as a unique (fundamental) cycle is created when we add a chord to a spanning tree, similarly a unique minimal cut is created when we add a spanning tree edge w to the set of chords (of the spanning tree). That is, removing all the chords of T leaves G connected, since the spanning tree T still remains. Then, if we further remove the spanning tree edge w, we separate the vertices of T (and G) into disconnected parts X and X^c, thus defining a cut. Furthermore, this is a minimal cut because if we leave any of its edges in G, the resulting graph is still connected (adding w would reconnect T; adding a chord would reconnect the parts of T disconnected by removing w). In analogy with fundamental cycles, we call such a minimal cut a *fundamental minimal cut* of G. The analog for fundamental minimal cuts of the representation theorem for fundamental cycles is given in the following theorem.

> **Theorem (Fundamental Minimal Cuts Representation).** Let $G(V, E)$ be a connected graph, and let T be a spanning tree of G. Then, every minimal cut m in G is the symmetric difference of the fundamental minimal cuts determined by the edges of m lying in T.

Cycles are symmetric differences of fundamental cycles, and minimal cuts are symmetric differences of fundamental minimal cuts; circuits are unions of edge disjoint cycles, and cuts are unions of edge disjoint minimal cuts. That is, we have the following.

> **Theorem (Circuit and Cut Representation).** Let $G(V, E)$ be a connected graph. Then, every circuit c in G is the union of edge disjoint cycles, and every cut m in G is the union of edge disjoint minimal cuts.

4-2 TREES AS MODELS

Trees and their invariants are used pervasively in the analysis of algorithms. Tree invariants such as height and path length determine the performance of search tree algorithms, while attempts to enhance the performance of search algorithms are often based on procedures that affect graph-theoretic invariants. For example, balancing algorithms minimize the height of binary search trees in order to enhance their performance. N-ary search trees (like B-trees) reduce tree height in exchange for increased vertex degree. Tree algorithms arise naturally in the compilation of algebraic expressions. Trees also occur as theoretical models of computation, such as the sort trees used to obtain lower bounds on the speed with which sorting can be done. Well-known techniques for optimal merging and data compression also use tree based algorithms. We shall consider some of these applications in the following.

4-2-1 Search Tree Performance

The performance of binary search trees can be analyzed in terms of the graph-theoretic properties of trees. For example, the height of a tree gives a worst case bound on the performance of a binary search, and we can estimate the average length of a binary search using the concept of the path length of a tree.

Let T be a binary search tree. When T is accessed for a key lying in T, the search terminates with a match at a vertex of T, which we will call an *internal vertex*. On the other hand, when T is accessed for a key not lying in T, the search terminates at a null pointer. For convenience, we will introduce special vertices, called *external vertices*, at each null pointer. The *internal path length* of T is then defined as the sum of the lengths of the paths from the root of T to its internal vertices, while the *external path length* of T is defined as the sum of the lengths of the paths from the root of T to its external vertices. The *average internal path length* of T equals the internal path length divided by the number of internal vertices, while the *average external path length* of T equals the external path length divided by the number of external vertices.

If each key (vertex) is weighted according to the frequency with which it is accessed, the sum of the lengths of all search paths from the root of T to all its internal vertices, weighted according to the frequency with which they are accessed, is called the *weighted internal path length* of T. If estimates are available for the frequencies with which each external vertex is accessed (corresponding to the frequency of a failed search for a given range of nonkeys), we can define the *weighted external path length* of T analogously. The optimal Huffman encoding technique described in a subsequent section, and the optimal search tree organization considered in Chapter 2, optimize the weighted path lengths of different tree models.

The performance of a binary search algorithm can be characterized in terms of the average internal and external path length of the tree. However, in order to model average behavior meaningfully, we need an appropriate model of randomness for trees. Since the trees we are interested in are generated by the standard binary tree insertion algorithm, the appropriate random model is to consider the trees produced by randomly permuting the possible keys $(1 .. n)$ and inserting them in permuted order, starting, of course, with an empty tree. Each permutation is equally likely and different permutations may lead to the same tree, so the likelihood of a given tree will be proportional to the number of permutations that give rise to the tree. Let us denote by $s(n)$ *the average number of comparisons needed by a successful key search* in a given random tree, and by $u(n)$ *the average number of comparisons needed by an unsuccessful key search*. Denote the internal path length for a given random tree with n keys by $I(n)$, and its external path lengths by $E(n)$.

There is a simple relationship between $I(n)$ and $E(n)$

$$E(n) = I(n) + 2n. \tag{4-1}$$

To prove this, we first observe that (4-1) is trivially true for a tree with one vertex. In general, when a new vertex v is added to an existing binary tree at distance x from the root of the tree, the internal path length is increased by x, while the external path length is decreased by x (because of the removal of the original external vertex) and then increased by $2x + 2$ (because of the addition of two new external vertices), for a net increase of $x + 2$. Thus, the difference between E and I is increased by 2 every time a vertex is added, from which (4-1) follows.

The following theorem establishes the average performance of binary search.

Theorem (Logarithmic Search Times). Let T be a random binary search tree of order n. Suppose the probability of access of an internal vertex (which contains a distinct integer key on $1 .. n$) is the same for each internal vertex, while the probability of access for each external vertex is the same for each external vertex. Then, $s(n)$ and $u(n)$ are both $O(\log n)$.

The proof is as follows. By assumption, any of the $n + 1$ possible types of failed searches are equally likely. Therefore, since the key we are accessing is (in our random model) equally likely to have been the first key, the second key, the third key, etc., inserted in the tree, and since the number of comparisons that are needed to find a key is one more than the number of comparisons needed when the key was first inserted, it follows that $s(n)$ satisfies:

$$s(n) = 1 + (u(0) + u(1) + \cdots + u(n - 1))/n. \tag{4-2}$$

There is a simple relation between $u(n)$ and $s(n)$ that will allow us to obtain a recurrence relation involving u (or s) alone. First, observe that by definition, s and u satisfy

$$s(n) = 1 + I(n)/n, \tag{4-3}$$

$$u(n) = E(n)/(n + 1), \tag{4-4}$$

from which it follows from (4-1) that

$$s(n) = ((n + 1)u(n)/n) - 1. \tag{4-5}$$

Combining (4-2) and (4-5), we obtain

$$(n + 1)u(n) = 2n + u(0) + \cdots + u(n - 1). \tag{4-6}$$

Subtracting (4-6) for $u(n)$ from (4-6) for $u(n + 1)$, we obtain

$$u(n + 1) = u(n) + 2/(n + 2), \tag{4-7}$$

from which it easily follows that $u(n)$ is $O(\log n)$. By (4-5), $s(n)$ is also $O(\log n)$. This completes the proof of the theorem.

A more precise result is given in Devroye (1986) who shows that the average height of a binary search tree on n vertices is asymptotic to $c \ln (n)$, where $c = 4.31107\ldots$. Flajolet and Odlyzko (1982) show that for rooted binary trees, where every vertex other than the root has either two children or is an endpoint, and where the tree is of "size" n, where n is the number of vertices in the tree with two children, the average height of the tree is $2(\pi n)^{(1/2)}$.

4-2-2 Abstract Models of Computation

Any sorting technique that sorts elements on the basis of a binary comparison of the (complete) sort keys is called a *comparison sort,* in contrast to a radix sort which makes sorting decisions on the basis of key fragments. Heapsort is a well-known comparison sort which sorts n elements in $O(n \log n)$ key comparisons and data movements. Using a theoretical model of comparison sorting which models the essential elements of any possible comparison sort, it is possible to show that no comparison can sort n elements in less than $O(n \log n)$ time.

Consider an arbitrary comparison sorting algorithm. The inputs to the algorithm are arrays of n elements whose values are the keys to be sorted. For convenience, we will assume that the keys are distinct. We may think of the function of the sorting algorithm as finding that unique permutation of any input array which sorts the array. If the elements are distinct, there are $n!$ possible sorting permutations, any one of which may be the sorting permutation for a particular input.

Theorem (Comparison Sort Lower Bound). Any comparison sort on n elements takes at least $O(n \log n)$ steps.

The proof is as follows. The sorting algorithm makes some sequence of data and algorithm dependent comparisons or decisions until the correct permutation has been identified (the array has been sorted). Though not explicit, the permutation is, nonetheless, always implicit in the operation of the algorithm. Thus, regardless of how the algorithm is implemented, we can interpret its operation as advancing through a binary decision tree. The internal vertices of the decision tree represent the binary decisions made by the algorithm. The endpoints of the tree represent the possible outcomes of an activation of the algorithm: one of the various possible sorting permutations. The size of the tree must be at least $n!$, since there are $n!$ endpoints in the tree. Consequently, the height of the tree must be at least $\log (n!)$. By Stirling's Formula, a well-known asymptotic approximation to $n!$, $\log (n!)$ is asymptotic to $O(n \log n)$. This completes the proof of the theorem.

4-2-3 Merge Trees

Data compression. An interesting application of trees is in the design of representations for compressing data. Suppose that $L = \{w_i, i = 1, \ldots, n\}$ is a list of words appearing in a text, and that word w_i has frequency f_i in the text. The words may be of fixed or varying length, or even single characters. If $len(w)$ denotes the length in bits of the usual representation for w, the length of the text (from which the frequency statistics were derived) is

$$\sum_{i=1}^{n} f_i \, len(w_i) \, . \tag{4-8}$$

For characters, the representations are usually fixed-length binary codes, while for words, the representations are typically of variable length but are independent of the frequency of the words; so that it may be possible to compress the text by assigning frequency dependent representations. If we represent the word w_i, $i = 1, \ldots, n$, by a bit string of length d_i, the length of the text using this representation becomes:

$$\sum_{i=1}^{n} f_i d_i \, . \tag{4-9}$$

We would like to determine a representation that minimizes the *compressed text length* (4-9).

The most obvious approach is to greedily represent the most frequent words in L with the shortest representations. For example, if $L = \{a, b, c\}$ with frequencies 3, 2, and 1 respectively, we could greedily represent a by the bit string 0, b by 1, and c by 00. However, the problem with this representation is that the encoded text cannot be unambiguously decoded. For example, does the bit string 00 represent the text *aa* or *c*? We can guarantee unique decodability if we require the representation to satisfy the *prefix property:* the representation assigned to any word must not match the leading part (prefix) of the representation assigned to any other word.

Any binary tree decoding scheme, in which the endpoints of the tree correspond to the words represented and a left (right) link in the path through the tree from the root to an endpoint corresponds to a 0 (1) bit in the representation of the word associated with the endpoint, defines a representation satisfying the prefix property. Refer, for example, to the binary tree in Figure 4-1b where an a is represented by the bit string 11, a b by the bit string 0, etc. Observe that the length d_i of the representation of a word w_i equals the distance of the endpoint corresponding to w_i from the root of the decoding tree. The words can be encoded initially using a lookup table whose entries give the word/representation correspondence.

There is a simple way to label a binary decoding tree that provides a useful way to interpret the cost (4-9) of a representation. Thus, suppose we label each endpoint of the tree with the frequency of its associated word; then label the parent of two endpoints with the sum of the labels of its children; and in general, for any internal vertex v (a vertex which is not an endpoint) label v with the sum of the labels of its children. Then, it is easy to see that the sum of the labels of the internal vertices equals the length (4-9) of the compressed message. Refer to Figure 4-2 for an illustration. Incidentally, (4-9) also equals the weighted internal path length of the decoding tree.

The following procedure constructs a binary decoding tree which minimizes the cost (4-9) for a set of words L of known frequency. The representation is called the *Huffman representation*.

(1) Sort the words in L in order of frequency. For each word w in L, create a rooted tree consisting of an isolated vertex labelled with the frequency of w. Denote the resulting set of trees by S.

Repeat steps (2) and (3) until S contains only one tree:

(2) Select two trees from S whose roots have the two smallest frequencies, call them f_1 and f_2, and form a new tree whose root has frequency $f_1 + f_2$, and has the roots of the selected trees as children.

(3) Remove the trees for f_1 and f_2 from S, and add the new tree of frequency $f_1 + f_2$ to S.

Character	Frequency	Representation
a	10	11
b	29	0
c	4	1011
d	9	100
e	5	1010

(a) Table of character frequencies and their Huffman codes.

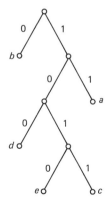

(b) Binary decoding tree. **Figure 4-1.** Huffman representation.

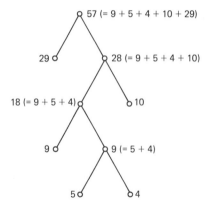

Figure 4-2. Binary decoding tree with internal labels shown.

Refer to Figures 4-1 to 4-3 for an example of the procedure.

The correctness and performance of the procedure are established by the following theorem.

Theorem (Performance and Optimality of Huffman Encoding). Let L be a list of n distinct words (of some uniformly bounded length b, independent of n), and suppose the frequency $f(w)$ of each word w in L is known with respect to a given text. Then, the Huffman encoding of the list can be determined in $O(n \log n)$ time, and the encoding minimizes the length of the compressed text.

The proof is as follows. Let $L = \{w_1, \ldots, w_n\}$ be the list of words, of frequencies f_i, $i = 1, \ldots, n$, respectively. Consider first the performance of the algorithm. Observe

(a) Frequencies: 4, 5, 9, 10, 29.

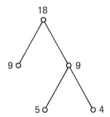

(b) Frequencies: 9, 9, 10, 29.

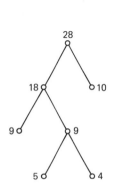

(c) Frequencies: 10, 18, 29.

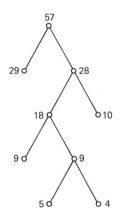

(d) Frequencies: 28, 29.

Figure 4-3. Example of Huffman algorithm.

that we can create a heap, ordered on word frequencies, in $O(n)$ time. (For a proof of this and the subsequent heap properties used here, refer to the section on minimum spanning trees.) Using a heap, it takes the algorithm $O(\log n)$ steps to find and remove the two least frequent words in L, which we can denote without loss of generality by f_1 and f_2. It then takes an additional $O(\log n)$ steps to insert the new value $f_1 + f_2$ in the heap. At this point, there are $n - 1$ values in the heap. We can assume by induction that it takes $O((n - 1)\log(n - 1))$ steps to create the Huffman tree on these frequencies. If we combine this with the preceding estimates, it follows that Huffman's algorithm can be implemented in $O(n \log n)$ time.

To establish the correctness of the procedure, we argue as follows. As we have observed, the cost of the encoding equals the sum of the weights of the internal vertices of the tree. Suppose f_1 and f_2 are the two smallest frequencies of the endpoints of the tree. We can assume that the endpoints for f_1 and f_2 are each at a maximum distance from the root of the tree; otherwise, we could reduce the weighted internal path length of the tree by exchanging, say, f_1 with a more distant endpoint (of necessarily greater frequency). We can also assume that f_1 and f_2 have the same parent; otherwise, we could exchange f_2 with the current sibling of f_1 again without increasing the weighted internal path length of the tree. Therefore, the parent vertex of f_1 and f_2, which must be of frequency $f_1 + f_2$, is an internal vertex of some optimal tree, which therefore has weight $W + (f_1 + f_2)$, where W equals the sum of the labels of the other internal vertices of that tree. But, these are also precisely the internal vertices of a tree on the endpoint frequencies: $f_1 + f_2, f_3, \ldots, f_n$, which we can assume by induction the algorithm finds correctly. Such a tree minimizes W, and so it also minimizes $W + (f_1 + f_2)$, the weight of the optimal tree on $f_1, \ldots, f_n$, as was to be shown.

Optimal file merging. The Huffman decoding tree has another application, to an apparently unrelated problem: How to merge a collection of sorted sequential files in such a way as to minimize the number of (file) records moved? For example, suppose we interpret the table in Figure 4-1a as referring to a list of files and interpret the frequencies in the table as referring to the number of records per file. Furthermore, rather than interpreting the tree in Figure 4-1b as a decoding tree, we think of it as defining the order in which to merge the files: Thus, first merge file-c and file-e; then merge the result of that merger with file-d; then merge that result with file-a; and finally merge that result with file-b. We define the total cost of the merger operations as the total number of records moved during the merging process. For example, in merging file-c and file-e, 9 ($= 4 + 5$) records are moved. In general, the total number of records moved depends both on the sizes of the files and the order in which they are merged. The cost is easily seen to equal the internal path length of the binary merging tree (where we consider the endpoint vertices as external vertices of the tree). Therefore, in order to minimize the cost, we merely construct a merge tree using the same procedure as for determining Huffman codes.

4-2-4 Precedence Trees for Multiprocessor Scheduling

The scheduling of tasks on a uniprocessor so as to optimize system performance parameters, like turn-around time, has been the subject of much investigation. It is far more complicated to properly schedule tasks in a multiprocessor environment, but there are some techniques available. A scheduling algorithm due to Hu (see Coffman and Denning (1973)) gives an optimal schedule under the following simplified circumstances. Suppose $T_1, \ldots, T_n$ are a set of tasks, each of which takes a unit time to execute. Suppose there are M homogeneous processors on which the tasks can be scheduled. Suppose also that the tasks are constrained by precedence relations defined by a rooted tree. Thus, if (T_i, T_j) is an edge of the precedence tree, task T_j cannot be started until task T_i has been completed. Figure 4-4a gives an example of a precedence tree. The following algorithm minimizes the completion time of such a set of tasks.

Hu's Scheduling Rule: Whenever a processor becomes free, start the next task all of whose precedent tasks have already completed *and* which has the highest level in the precedence tree of all tasks that have not yet been started.

Figure 4-4 illustrates the application of the algorithm to a set of 12 tasks, under the constraints of the precedence tree shown in Figure 4-4a and with $M = 3$ processors available. Figure 4-4b shows the successively scheduled sets of tasks. Figure 4-4c gives the *Gannt chart* for the schedule. The proof of trees of correctness of this straightforward algorithm is nontrivial.

4-3 MINIMUM SPANNING TREES

The minimum spanning tree problem is a classic situation where the greedy method yields an optimal solution. It also illustrates how the careful design of data structures can substantially improve the performance of an algorithm. We will use heaps and dis-

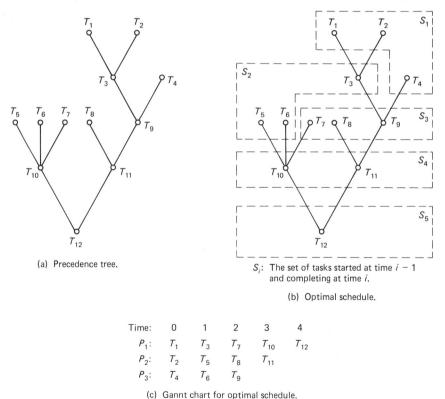

(a) Precedence tree.

S_i: The set of tasks started at time $i - 1$ and completing at time i.

(b) Optimal schedule.

Time:	0	1	2	3	4
P_1:	T_1	T_3	T_7	T_{10}	T_{12}
P_2:	T_2	T_5	T_8	T_{11}	
P_3:	T_4	T_6	T_9		

(c) Gannt chart for optimal schedule.

Figure 4-4. Example of Hu's optimal scheduling algorithm.

joint sets represented as trees to implement a fast version of the algorithm. The analysis of the performance of these data structures is also instructive. The analysis of the disjoint sets uses a classical tree-height versus tree-size argument, while the analysis of heap creation illustrates the application of recurrence relations in performance analysis.

We define the minimum spanning tree problem as follows. Given an undirected graph G with real-valued edge weights assigned to each of its edges, find a spanning tree T of G that has minimum total edge weight. The problem has a simple interpretation. Consider the complete graph $K(n)$, where each vertex corresponds to a city and the weight of an edge corresponds to the cost of establishing a direct communication link between the pair of cities corresponding to the endpoints of the edge. A minimum spanning tree on $K(n)$ corresponds to a least cost communication network connecting all the cities.

The optimal solution to the minimum spanning tree problem is surprisingly simple. We merely construct a tree, starting with an empty tree, iteratively adding edges to the tree in increasing order of edge weight, excluding an edge only if it forms a cycle with the previously added edges. We terminate the process when a spanning tree is formed. Figures 4-5 and 4-6 illustrate the idea. The algorithm is greedy because it optimizes the possible improvement in the objective function, here the weight of the spanning subgraph, at each successive step of the algorithm, subject only to the constraint that no cycles are formed.

Minimum spanning tree algorithm.   We will initially describe the algorithm from a high-level point of view. Later, we will give a more detailed version using the heap and disjoint set data structures to be developed. The data type of the graph is

```
type   Graph = record
               |V|:  Integer
               |E|:  Integer
               Edges(|E|,2): 1..|V|
               Weight(|E|): Real
        end
```

The $|E| \times 2$ array **Edges** contains the edge list for G. The ith edge (u, v) is represented by the ith row of the array **Edges** (that is, (**Edges(i, 1)**, **Edges(i, 2)**)), where **Edges(i, 1)** equals u and **Edges(i, 2)** equals v, for $i = 1 .. |E|$. Weight(i) gives the

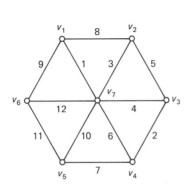

Figure 4-5.   A weighted graph G.

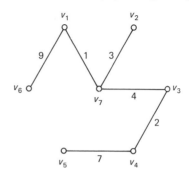

Figure 4-6.   Minimum spanning tree on G.

length of the ith edge, $i = 1 \ldots |E|$. For convenience, we also use a type Edge_Entry defined as

```
type  Edge_Entry = record
                x₁, x₂: 1..|V|
                xweight: Real
          end
```

We assume that G is connected and has an order of at least 2. If H is a graph and (u, v) is an edge, $H \cup (u, v)$ has the natural interpretation that if either vertex u or v is not in $V(H)$, it is (they are) added to the set of vertices; then the edge is added. The statement of the algorithm is as follows.

```
Procedure MST(G,T)

(* Returns a minimum spanning tree for G in T *)

var   G, T: Graph
         x: Edge_Entry

Set |V(T)| to |V(G)|
Set |E(T)| to 0

repeat

      Select the next smallest edge x in E(G) − E(T)

      if  T ∪ x  is acyclic  then  Set T to T ∪ x

until  |E(T)| = |V(G)| − 1

End_Procedure_MST
```

Before proceeding to a refinement of the algorithm, we will establish its correctness.

Theorem (Correctness of Minimum Spanning Tree Algorithm). Let $G(V, E)$ be a connected weighted graph, $|V| > 1$, and let T be the weighted spanning tree of G constructed by the MST algorithm. Then, T is a minimum spanning tree of G.

The proof is as follows. First, observe that the subgraph T constructed by the algorithm is certainly a spanning tree. Then, let S be an arbitrary minimum spanning tree in G. We will show S and T have the same total edge weight; so T must also be a minimum spanning tree of G. Our approach will be to iteratively add and remove edges from S, without increasing its total edge weight, but increasing the number of edges it has in common with T, until eventually it becomes identical to T.

If T and S are not identical, there is some edge in T which is not in S, and conversely. Let x be the first edge the algorithm adds to T which is not also in S. Since S is a spanning tree, x must be a chord of S, and so $S \cup x$ must contain a cycle C. Since T is acyclic, there must exist some edge x' on C which lies in S but not in T. We will show the spanning tree $S \cup x - x'$ weighs no more than S.

We will derive a contradiction from the assumption that x' weighs less than x. If x' weighs less than x, then x' must have been considered by the algorithm for inclusion in T before x was considered. But, by the definition of x, every edge in T added prior to x also lies in S. Since x' is not in T, the subgraph of T formed by the minimum spanning tree algorithm up to the point when x' was examined together with x', must have contained a cycle C'. But, all the edges on C', including x', lie in S, which is acyclic. Therefore, x' could not have caused a cycle in T. Therefore, the algorithm would have added x' to T, contrary to our assumption that x' is not in T. It follows that x' must weigh at least as much as x.

Therefore, the tree $S \cup x - x'$ weighs no more than S and so is a minimum spanning tree. This tree also has one more edge in common with T than S does. Consequently, if we iterate this process, we will eventually arrive at a minimum spanning tree which is identical to T. This completes the proof.

Preliminary performance analysis.   We will give a rudimentary analysis of the performance of the algorithm in order to identify bottlenecks where special data structures might prove advantageous.

First, observe that although there are only $|V| - 1$ edges in the MST tree, we may have to examine as many as $|E|$ edges in order to determine which edges belong in T. Therefore, the **repeat** loop in the algorithm may be executed as many as $|E|$ times. At each execution of the loop body, we select the next least weight edge from a shrinking list of edges of cardinality as large as $O(|E|)$. If the edges are sorted by weight, which takes (preprocessing) time $O(|E| \log |E|)$, each edge selection can be done in $O(1)$ time.

Each execution of the loop also tests whether adding an edge introduces a cycle. Because the underlying graph T is always a forest, this test can be done easily. At any point, T is acyclic and so consists of a collection of components which are trees. Therefore, adding an edge to this forest can introduce a cycle only if both the endpoints of the edge lie in the same component of the forest. Each component of the forest comprises a set of vertices, and so we can represent the forest, at least for testing acyclicity, as a collection of disjoint sets. To determine whether adding an edge introduces a cycle, we need only test whether both the endpoints of the edge lie in the same disjoint set. Since these sets can be $O(|V|)$ in size, this test can take $O(|V|)$ time.

This preliminary analysis suggests the complexity of the algorithm is $O(|V||E| + |E| \log |E|)$, which is just $O(|V||E|)$. Of course, we can drastically improve this performance by choosing suitable data structures for the two critical steps of the algorithm.

Efficient minimum selection.   It takes time $O(|E| \log |E|)$, or equivalently $O(|E| \log |V|)$, to sort the edges by weight. Since we may examine as few as $|V| - 1$ edges, sorting all $|E|$ edges seems excessive. But, how can we improve on this? If we use ordinary selection to get the next smallest edge in $O(|E|)$ time and repeat this M times (in the case where the tree is constructed after examining M edges), the cost of this part of the algorithm is $O(M |E|)$, which is inferior to fast complete sorting since M is at least $|V| - 1$. Moreover, the overall performance of the algorithm will improve only if faster edge selection is matched by faster cycle testing. With this in mind, we will now consider the advantages of a heap for minimum selection.

We will assume familiarity with the definition of a heap. Recall that a *heap* is a balanced binary tree, ordered on sort keys located at each of its nodes. A heap is usually represented by an array $H(1 . . N)$, where $H(i)$ acts as the parent of $H(2i)$ and $H(2i + 1)$. The basic *heap operations* are

(H1) Create (a heap),

(H2) Find_min (find the minimum element in the heap),

(H3) Delete (the least element in the heap and restore the heap),

(H4) Insert (a new element into a heap),

(H5) Member (test if an element is in a heap), and

(H6) Change (the value and if necessary the position of an existing heap element).

We have already described the Member and Change operations under Dijkstra's algorithm. The MST algorithm uses only the Create and Delete operations. We will review the Delete, Create, and Insert operations. We use the following type definition, which will be slightly modified later when we deal with the edge heaps needed for the MST algorithm.

```
type   Heap = record
                H(N): Integer
                Last: 0..|E|
              end
```

The Delete procedure follows. For an $|E|$ edge heap, the procedure has performance $O(\log |E|)$.

```
Function Delete (A, Small)

(* Returns the least heap element in Small and deletes it, or
   fails *)

var   A: Heap
      I, Small: Integer
      Delete: Boolean function

if    Last = 0   then   Set Delete to False
                        return

Set   Small to H(1)
Set   H(1)  to H(Last)
Set   Last   to Last − 1
Set   Delete to True

(* Sift root element into correct heap position *)

Set I to 1

while   Last ≥ 2I   do   Swap H(I) with its smaller child
                         Set I to position of child

End_Function_Delete
```

Insert adds an element to the heap. We implement it by adding the new element at H(Last + 1) and sifting it upwards as long as its value is smaller than the value of its parent. Insert also takes $O(\log |E|)$ time.

Create Algorithm and Its Performance. Creating a heap by iterating the Update procedure takes $O(|E| \log |E|)$ time, since there are $|E|$ updates and each update takes $O(\log |E|)$ time. But, this is just the time it takes to sort $|E|$ edges by weight, which we are attempting to avoid. Actually, we can Create a heap in $O(|E|)$ time, as we shall now show.

A heap may be defined recursively as consisting of an element Root containing the smallest value in the heap, together with subtrees rooted at the left and right children of Root which are also heaps. This definition suggests a recursive procedure for creating a heap. We will assume for simplicity that the heap has $|E|$ equal to $2^k - 1$ elements, for some positive integer k.

The Create procedure is as follows.

(1) Divide the set of elements to be represented in the heap into three sets of size 1, $2^{k-1} - 1$, and $2^{k-1} - 1$.

(2) Recursively construct subheaps on each of the two sets of size $2^{k-1} - 1$.

(3) Make the subheaps created in (2) the left and right subheaps of the remaining element.

(4) Sift the root of the structure in (3) into its correct heap position using the same technique as in the Delete algorithm.

Upon completion, the smallest element of the data structure lies at the root, and the left and right subtrees of the root are heaps. Therefore, the overall structure is a heap. The performance of this procedure can be analyzed using recurrence relations. The analysis is a nice illustration of how performance bounds can be obtained from the recurrence relations determined by recursive procedures. We summarize the analysis in the following theorem.

Theorem (Linear Time Heap Creation). Let X be a set of N real numbers. Then, a heap on X can be created in $O(N)$ steps.

The proof is as follows. Let W denote the amount of work required to create a heap on X using the recursive algorithm described. We can consider W as a function either of the cardinality N of X or the height of the heap. If we consider W as a function of N, we obtain the recurrence relation

$$W(N) = 2W(N/2) + c \log(N).$$

This reflects the recursive construction of the heap from two subheaps of size $N/2$, and the subsequent $O(\log N)$ steps required to sift the root of the heap into its proper position. On the other hand, if we consider W as a function of the height H of the heap created, the recurrence relation is

$$W(H) = 2W(H - 1) + cH.$$

We will work with the recurrence in this form, since it is slightly more convenient to handle. If we feed this recurrence back into itself repeatedly, we obtain

$$W(H) = 2^{H-1}W(1) + c(H + 2(H - 1) + \cdots + 2^i(H - i) + \cdots + 2^{H-1}(1)),$$

where c is a constant. The leading term in this expression is $O(N)$, since N is at most $2^H - 1$. The remainder of the expression must be summed carefully. If we replace $H - i$ by j and set r equal to $1/2$, then the summation becomes

$$2^H \sum_{j=1}^{H} jr^j .$$

Since the factor 2^H is $O(N)$ and the summation is bounded, this term is also $O(N)$. This completes the proof of the theorem.

Summarizing, we can create an $|E|$ element edge heap in $O(|E|)$ time, and delete an element from the same heap in $O(\log |E|)$ time. Therefore, if the MST algorithm uses M heap operations, then we can perform them in $O(|E| + M \log |E|)$ time. In the case that M is $O(|V|)$, this is $O(|E| + |V| \log |E|)$ steps; while if M is $O(|E|)$, it is $O(|E| \log |E|)$ or equivalently $O(|E| \log |V|)$ steps.

Efficient acyclic test. We will now show how to test whether an edge forms a cycle in $O(\log |V|)$ time. As we have observed, the forest constructed by the MST algorithm consists of components. Each component is a tree comprising a set of disjoint vertices. We can represent each disjoint set by a unary tree. The *unary tree* representation of a set S is a tree with one node for each member of the set. Each node has a unique parent node in the unary tree which the node points to, except for the root of the unary tree which points to itself. We consider the root of the unary tree as the representative or owner of the set represented by the tree. We shall introduce two unary tree functions

(1) Root (UT, rootx, x), and

(2) Union (UT, rootx, rooty),

where UT is a (collection of) unary tree(s). Root (UT, rootx, x) returns the root of the tree representing the set containing x in both Root and rootx. Union(UT, rootx, rooty) forms the union of the pair of disjoint sets whose unary trees have roots rootx and rooty.

Root (UT, rootx, x) is trivial to implement. We merely follow the pointers to parent nodes, starting with x, all the way to the root of the tree containing x. We can implement Union(UT, rootx, rooty) by merely making rootx point to rooty. The performance of the Root operation depends on the heights of the trees, so it is important to keep the tree heights small. The following rule ensures this.

Small-large rule: When forming a union of two unary trees, always make the root of the tree of smaller cardinality point to the root of the tree of larger cardinality.

To apply the small-large rule we need a field (Size) in the root of each unary tree which contains the number of vertices in the tree. Then, whenever we form the union of two trees, we update the Size field for the root of the union.

The performance of Union (UT, rootx, rooty) is $O(1)$ since it merely requires setting rootx to point to rooty, or vice versa. On the other hand, Root (UT, rootx, x) takes an amount of time proportional to the height of the unary tree that represents the set x belongs to. But, this in turn depends critically on how we implement Union. We can show that the heights of the unary trees that result if we follow the small-large rule are always logarithmic in their size, giving $O(\log |E|)$ performance for Root (for a collection of unary trees of total cardinality $|E|$). In the following theorem Height(UT) denotes the height of a unary tree and $|UT|$ denotes its number of vertices.

Theorem (Logarithmic Height Unary Trees). Let M be a set of disjoint singletons. Let S be a subset of M constructed by applying a sequence of Union operations to the members of X and following the small-large rule. If $T(S)$ denotes the unary tree representing S, then $|T(S)|$ and Height($T(S)$) satisfy

$$\text{Height}(T(S)) \leq \lg|T(S)|\,.$$

The proof of the theorem is as follows. For convenience, we shall work with the equivalent relation,

$$\text{If Height}(T(S)) \geq h \quad \text{then } |T(S)| \geq 2^h\,.$$

The proof is by induction on h. Assume that every unary tree of height $h - 1$, has size at least 2^{h-1}. We shall show that if T is a least cardinality tree of height h constructed by a sequence of operations of the type described in the theorem, T has at least 2^h vertices. It will then follow, by the definition of T, that every other tree of height h has at least that many vertices, completing the induction.

By supposition, T is the union of a pair of trees X and Y, whose sizes are less than $|T|$. Therefore, since T is the smallest tree of height h, X and Y must each have height less than h. Suppose without loss of generality that $|Y| \leq |X|$. Therefore, we can assume that Y was appended to the root of X. Height(X) $\leq h - 1$. If Height(Y) $\leq h - 2$, the height of their union, namely, Height(T) $\leq h - 1$, contrary to supposition. Therefore, Height(Y) must equal $h - 1$. It follows by induction that $|Y| \geq 2^{h-1}$. Since $|X| \geq |Y|$, then $|X| \geq 2^{h-1}$ also. Therefore, T must have at least 2^h vertices, as was to be shown. This completes the proof of the theorem.

All the disjoint sets considered in the MST algorithm are initially singletons: corresponding to the isolated vertices of a spanning forest with no edges. Figure 4-7 illustrates the sequence of disjoint sets obtained for the MST algorithm, following the small-large rule, for the example of Figures 4-5 and 4-6.

The number M of acyclicity tests required by the MST algorithm depends on the graph G being processed. M acyclicity tests take time $O(M \log |V|)$. If M is $O(|E|)$, this is comparable to the time for the worst case heap requirements. If M is $O(|V|)$, the operations take time $O(|V| \log|V|)$. If we combine this with the corresponding $O(|E| + M \log |E|)$ time for the heap operations, we obtain an overall $O(|E| + M \log |E|)$ time for the algorithm. Even in the worst case, the enhanced algorithm only takes $O(|E| \log|V|)$ time, while the unenhanced version takes time $O(|E||V|)$. Thus, the new data structure will always improve the performance of the algorithm, though the improvement is a function of the problem dependent parameter M.

Enhanced minimum spanning tree algorithm. We will now give another implementation of the minimum spanning tree algorithm using the enhanced heap and disjoint set data structures we have introduced. The heap and unary tree data types are defined as follows.

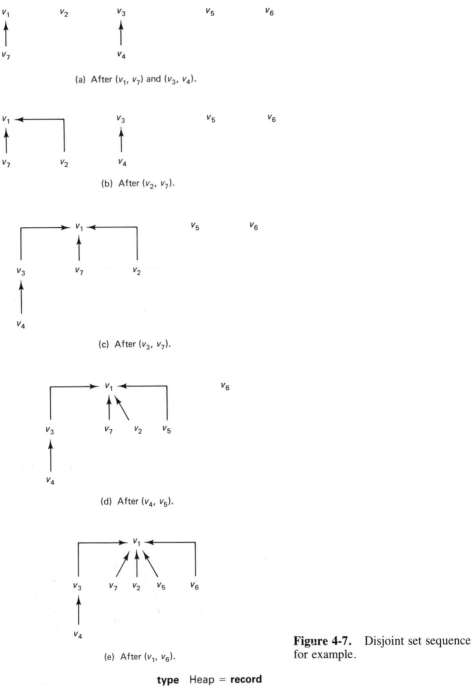

(a) After (v_1, v_7) and (v_3, v_4).

(b) After (v_2, v_7).

(c) After (v_3, v_7).

(d) After (v_4, v_5).

Figure 4-7. Disjoint set sequence for example.

(e) After (v_1, v_6).

type Heap = **record**
    H(|E|,2): 1..|V|
    Weight(|E|): Real
    Last: 0..|E|
end

The edges of G, in heap order with respect to their weight, are stored in H. Their weights are stored in the same order in Weight. Last gives the index of the last element in the heap. Whenever we interchange heap elements in the process of maintaining the heap, we interchange both the entries in H and Weight.

The Unary Tree representation for disjoint sets is

```
type  Unary Tree  = record
                      Parent(|V|): 0..|V|
                      Size(|V|):   0..|V|
                    end
```

The pointers of the unary tree are stored in Parent. Initially, Parent(i) = i and Size(i) = 1, for $i = 1 .. |V|$. Generally, if vertex i is the owner (root) of a component, Size(i) equals the number of vertices in its component.

The enhanced MST algorithm follows. It uses the utility Create_Heap (G_Edge, G) to store the edges of G in the heap G_Edge. Create_Unary (T_Vert, G) initializes the disjoint set representation for the vertices of G in T_Vert. Initial_Graph(T, G) initializes the tree T using $|V(G)|$. Add(T, x) adds an edge x to T. The type Edge_Entry is as defined previously.

```
Procedure MST (G, T)

(* Returns a Minimum Spanning Tree for G in T *)

var  G, T: Graph
     G_Edge: Edge Heap
     T_Vert: Unary Tree
     x: Edge_Entry
     Rootx₁, Rootx₂: 1..|V|
     Root: Integer function

Create_Heap (G_Edge, G)
Create_Unary (T_Vert, G)
Initial_Graph (T, G)

repeat

  Delete (G_Edge, x)

  if    Root (T_Vert, Rootx₁, x₁) <> Root (T_Vert, Rootx₂, x₂)

  then  Union (T_Vert, Rootx₁, Rootx₂)
        Add (T, x)

  until  |E(T)| = |V(G)| − 1

End_Procedure_MST
```

Path compression.   Path Compression is a technique which can be used to improve the performance of a unary tree reresentation, and works as follows. After we find the root u of the tree containing a given node v, we make all the nodes along the unary tree path from v to u point directly to u. That is, for each node w on the path from v to u, we make u the parent of w. We shall also introduce a variation of the small-large rule where the so-called ranks of the roots of the unary trees are used for the small-large comparisons. (The term rank used here is not the same as the ranks to be defined in Section 4-6.) The ranks are initialized to 0 for each trivial root. Then, whenever a union is formed, the root of smaller rank is linked to the root of larger rank. The rank is incremented by 1 whenever a pair of trees of equal rank are joined; otherwise, the rank of the new root remains the same. If we combine both path compression and this version of the small-large rule, we improve the unary tree performance significantly. The performance bound is in terms of the inverse function of a very rapidly growing function called the Ackerman function $A(i, j)$, defined for $i, j \geq 1$: $A(1, j) = 2^j$ for $j \geq 1$, $A(i, 1) = A(i-1, 2)$ for $i \geq 2$, and $A(i, j) = A(i-1, A(i, j-1))$ for $i, j \geq 2$. The function $a(m, n)$, for $m \geq n \geq 1$, is then defined by $a(m, n) = \min \{i \geq 1 \mid A(i, G(m/n)) > \log n\}$, where $G(m/n)$ denotes the greatest integer not larger than m/n. Practically speaking, $a(m, n)$ is not larger than 4.

Theorem (Performance under Path Compression and Small-large Rank Rule). An intermixed sequence of m Root and Union operations on a unary tree on n elements takes time $O(ma(m, n))$. (We assume the tree is started in its trivial state on n nodes.)

The proof of this theorem is quite difficult and is discussed in Tarjan (1983) (other references at the end of the chapter). It follows from the theorem that if the edges of the graph are already sorted by edge weight, a minimum spanning tree can be found in $O(|E| a(|V|, |E|))$ steps. For an even faster, but more complicated, minimum spanning tree algorithm, see Section 4-6.

4-4 GEOMETRIC MINIMUM SPANNING TREES

The preceding section established an $O(|E| \log |V|)$ algorithm for finding a minimum spanning tree. We will now show how to use the Voronoi diagrams introduced in Chapter 2 to find a minimum spanning tree of a graph in the plane in $O(|V| \log |V|)$ steps, a substantial improvement over the previous algorithm. Assume V is a set of points in euclidean 2-space. Let $G(V, E)$ be the induced complete euclidean graph on the points of V, where the edges of $G(V, E)$ correspond to the euclidean line segments between the points of V. Recall that although G is embedded in the plane, it need not be a planar graph, since its edges can intersect at points other than the points (vertices) of V. We will show how to find a minimum spanning tree of G using Voronoi diagrams in three steps: (1) we define a planar (not merely euclidean) graph associated with the Voronoi diagram $\text{Vor}(V)$, and denoted by $\text{Dual}(V)$; (2) we find a minimum spanning tree of $\text{Dual}(V)$, in $O(|V| \log |V|)$ time; and (3) we show a minimum spanning tree of $\text{Dual}(V)$ is necessarily a minimum spanning tree of $G(V, E)$.

The associated graph $\text{Dual}(V)$ is defined as follows. The set of vertices of $\text{Dual}(V)$, $V(\text{Dual}(V))$, equals V. Vertices v_i and v_j in $V(\text{Dual}(V))$ are adjacent if and only if the Voronoi polygons $\text{Vor}(v_i, V)$ and $\text{Vor}(v_j, V)$ of v_i and v_j share an edge of $\text{Vor}(V)$. $\text{Dual}(V)$ is a subgraph of $G(V, E)$, and can be easily constructed from $\text{Vor}(V)$ in $O(|V|)$ time using its definition. Since $\text{Dual}(V)$ is a planar graph, it has only $O(|V|)$ edges, so the MST algorithm can find a minimum spanning tree of $\text{Dual}(V)$ in $O(|V| \log |V|)$ time.

To prove that a minimum spanning tree of Dual(V) is automatically a minimum spanning tree of V, we use an alternative version of the MST algorithm devised by Prim. While the previous MST algorithm constructed a minimum spanning tree by greedily merging the component trees of a spanning forest, Prim's algorithm, in contrast, constructs a minimum spanning tree by expanding a (nonspanning) subtree in the greedy manner described by the following high level algorithm.

```
Procedure PRIM_MST(G,T)

(* Returns a minimum spanning tree for G in T *)

var    G,T: Graph
       v,v₁: Vertex
         x: Edge

Set T to an arbitrary vertex v₁ in V(G)

repeat   |V(G)| − 1   times

    Select the least weight x (= (u,v)) in E(G)
    such that u is in V(T) and v is not in V(T).

    Add x to T.

End_Procedure_PRIM_MST
```

This algorithm clearly constructs a spanning tree, which we leave it as an exercise to show is a minimum spanning tree. It then follows from the lemma below that a minimum spanning tree of Dual(V) is also a minimum spanning tree of $G(V, E)$, as was to be shown.

Lemma. The tree constructed by Prim's algorithm lies in Dual(V).

The proof is as follows. The idea is to follow the algorithm in a step by step fashion, interpreting its operation in terms of the corresponding Voronoi diagram Vor(V). Thus, let (v_1, t) be the least weight edge of G which is incident with v_1. The vertex t belongs to the Voronoi polygon Vor(t, V) which borders Vor(v_1, V). Therefore, the edge (v_1, t) must be an edge of Dual(V). Consequently, the first edge added by the algorithm to T is an edge of the Dual(V). Similarly, at the next step, Prim's algorithm seeks the least weight edge of the form (v_1, u) or (t, u). By the definition of the Voronoi diagram, u must be a Voronoi neighbor of Vor(v_1, V) $\cup$ Vor(t, V). Therefore, once again, whichever edge the algorithm adds to T is an edge of Dual(V). The general case follows the same idea. That is, the algorithm, as it proceeds, merely adds an edge of the dual of Vor(V) to T at each step. Since all the edges of the spanning tree constructed by the algorithm lie in the dual, the final spanning tree must be a subgraph of Dual(V), as was to be shown.

4-5 ACYCLIC DIGRAPHS

This section considers several applications of acyclic digraphs and the graphical properties they are based on: parts explosions and topological orders, deadlock avoidance and cycle detection, scheduling tasks under prerequisite constraints and longest paths, and register allocation for expression evaluation and digraph labelling.

4-5-1 Bill of Materials (Topological Sorting)

A *parts explosion* is a method of describing the hierarchical structure of a composite product in terms of its components. An example is shown in Figure 4-8a where an acyclic digraph $G(V,E)$ is used to model a parts explosion. Each vertex in the digraph corresponds to a part of the product or a subassembly of parts of the product and is labelled with the name of the part. There is an edge (x, y) if the part or subassembly y is a component of the assembly x. The edge is labelled with the number of copies of the part (subassembly) used in the parent part (assembly). A primitive part is by definition one containing no components. It is represented in the digraph model by a vertex of out-degree zero. The part represented by the parts explosion is called the root part. A *bill of materials matrix* **M** for the parts explosion is shown in Figure 4-8b. Component **M(i, j)** gives the multiplicity of j as a component of i.

We consider how to compute the number of primitive parts required to make a single composite part. The composite part has the nominal requirements list $P = (p_1, \ldots, p_n)$, where p_i gives the number of parts of type i required. Let k be the length of the longest path in the acyclic parts explosion digraph. The number of primitive parts required to realize P are determined by the following procedure. The final value returned in P is the list of primitive parts requirements. Refer to Figure 4-8c.

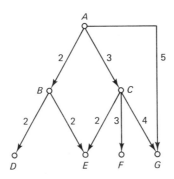

(a) Acyclic diagram for parts explosion.

	A	B	C	D	E	F	G
A	0	2	3	0	0	0	5
B	0	0	0	2	2	0	0
C	0	0	0	0	2	3	4
D	0	0	0	1	0	0	0
E	0	0	0	0	1	0	0
F	0	0	0	0	0	1	0
G	0	0	0	0	0	0	1

(b) Bill of materials matrix **M**.

	A	B	C	D	E	F	G
P:	1	0	0	0	0	0	0
PM:	0	2	3	0	0	0	5
(PM)M:	0	0	0	4	10	9	17

(c) Component requirements for one A part: (PM)M.

Figure 4-8. Bill of materials calculation.

Procedure Parts_Requirement (M, P, k)

(* Returns, for the input parts requirements list P, the
    primitive parts requirements list, also in P *)

var n: Integer constant
        M(n, n), P(n), i, k: Integer

for i = 1 to k **do** **Set** P to P M

End_Procedure_Parts_Requirement

We have assumed the parts are indexed so that the bill of materials matrix is in upper-triangular form. This can be used to simplify both the matrix calculations and storage requirements. The matrix is automatically upper triangular if the vertices of the model digraph $G(V, E)$ are indexed, from $1 .. |V|$, in such a way that every vertex has a distinct index, and, if (u, v) is an edge, the index assigned to u is smaller than the index assigned to v. Such an ordering of the vertices is called a *topological ordering*. The vertices of an acyclic digraph can always be topologically ordered.

Theorem (Topological Ordering).   The vertices of a digraph $G(V, E)$ can be ranked in topological order if and only if G is acyclic.

The proof is simple. The only if part is obvious. To prove sufficiency, observe that every digraph necessarily has at least one vertex of out-degree zero, such as one obtained by repeatedly extending any path until it becomes blocked. Similarly, there also exists some vertex v of in-degree zero. If we set Index(v) to 1, and repeat the same process on the acyclic subgraph $G - v$, each time incrementing the index assigned to the vertex of in-degree zero, eventually all of G will be indexed. By construction, whenever (u, v) is an edge of G, u must be indexed before v, so that Index(u) must be less than Index(v). This completes the proof.

A procedure for topologically numbering a digraph is as follows. We scan the adjacency list of the graph to determine the in-degrees of every vertex, in $O(|V| + |E|)$ time, and then create a list of the vertices of in-degree zero, denoted the zero-list, in $O(|V|)$ time. We then initialize a counter k to 0, select a vertex from the zero-list, label it with k (and increment k), and then decrement the in-degrees of every vertex on its adjacency list. Any vertices whose in-degrees are thereby reduced to zero, are also placed on the zero-list. We repeat this process until the zero-list is empty, which takes time $O(|E|)$. If any vertices remain that were not put on the zero-list, G is not acyclic; otherwise G is acyclic and the labelling is a topological ordering of its vertices.

A recursive procedure Topological_Order for topologically indexing the vertices of an acyclic digraph is given below. From the viewpoint of Chapter 5, Topological_Order (G) can be considered as an enhanced depth first search. The procedure assumes the digraph is acyclic, and has $O(|V| + |E|)$ performance, since each vertex and edge is visited only once. $G(V, E)$ is represented by a linear array, wherein each vertex has a field, Index, where the topological rank of the vertex is stored. The procedure uses an auto-decrementing function Subtract_count which is initialized to $|V|$ and decremented by 1 after each invocation.

Procedure Topological_Order (G)

(* Topologically order G *)

```
var   G: Graph
      v: 1..|V|
      Subtract_count: Integer function

for every v ∈ G   do   Set Index(v) to 0

for every v ∈ G   do   if Index(v) = 0
                       then Topological_Order (G,v)


End_Procedure_Topological_Order

Procedure Topological_Order (G,v)

(* Rank the unnumbered vertices in G that are reachable from v in
   topological order *)

var   G: Graph
      v, w: 1..|V|

for every w ∈ ADJ(v) do if     Index(w) = 0
                        then   Topological_Order(G, w)

Set Index(v) to   Subtract_count

End_Procedure_Topological_Order
```

4-5-2 Deadlock Avoidance (Cycle Testing)

Deadlock is a phenomenon of resource sharing arising when limited resources cannot be simultaneously shared. It occurs when the processes using a set of resources become embroiled in a vicious cycle of resource requests and allocations. As an example, consider the case where two processes *A* and *B*, each of which may require up to three tape drives, share access to four tape drives. If *A* and *B* are each initially allocated two drives and *A* requests a third drive, *A* must be suspended (blocked) until *B* releases one of the drives already allocated to it. If at this point *B* also happens to request another drive, *B* is also blocked. Now, both processes are deadlocked, since each is waiting for the release of a resource allocated to the other suspended process, which de facto cannot release the resource.

Both of the following conditions must be satisfied for a deadlock to be possible:

(1) *Mutual Exclusion,* that is, at least one of the resources must be nonshareable in the sense that only one process at a time can use the resource, and

(2) *No Preemption,* that is, the unshareable resource(s) in (1) cannot be preempted; that is, the resource can only be released by the process to which it has been allocated only after that process has completed using the resource.

We shall make several assumptions about the environment: there is only a single instance of each type of resource; each process in the system lists all the resources it may possibly require prior to executing; and every process starts executing simultaneously.

A method of allocating resources in such a way as to prevent deadlock using a *resource allocation digraph* $G(V, E)$ is as follows. $G(V, E)$ has two types of vertices, resource and process vertices, and so is bipartite. There is a resource vertex for each instance of a resource, and a process vertex for each process. There are three types of edges, *claim edges*, *request edges*, and *assignment edges*.

We add, delete, or change edges depending on the requests, allocations, and deallocations made according to the following rules:

(1) If a process p may require a resource r, we enter a claim edge (p, r) in G, when the system is initialized.

(2) If a process p requests a resource r and r is unavailable, we change the claim edge (p, r) to a request edge (p, r).

(3) If a process p requests a resource r and r is available, we remove the claim edge (p, r) and enter an assignment edge (r, p).

(4) If a process p releases a resource r, we remove the assignment edge (r, p) and enter a claim edge (p, r).

(5) If there is a process q blocked on resource r when r is released, we allocate r to q, enter the assignment edge (r, q), and remove the request edge (q, r).

The resource allocation is subject to the *deadlock avoidance rule:* allocate a resource r to a process p only if allocating r does not cause a cycle in G.

The required cycle testing can be done, for example, using depth first search (refer to Chapter 5), or using more efficient techniques based on the special structure of the problem (namely, the graph is bipartite and existing edges are merely reversed). An example is shown in Figure 4-9. Figure 4-9a gives the current state of the system. If p_2 requests r_2, then r_2 should not be allocated to p_2 even though r_2 is currently available, by virtue of the deadlock avoidance rule. Otherwise, the state in Figure 4-9b will occur. Although that state need not lead to a deadlock, we cannot guarantee at this point that a deadlock will not occur. For example, if p_1 subsequently requests r_2 before p_1 releases r_1, deadlock will occur because p_1 will be blocked waiting for r_2, while p_2 is blocked waiting for r_1.

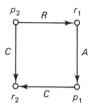 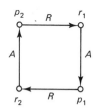

Resources	r_1, r_2
Processes	p_1, p_2
Claims	$(p_1, r_2), (p_2, r_2)$
Requests	(p_2, r_1)
Assignments	(r_1, p_1)

(a) State of system before p_2 requests r_2: no cycle.

Resources	r_1, r_2
Processes	p_1, p_2
Claims	(p_1, r_2)
Requests	(p_2, r_1)
Assignments	$(r_1, p_1), (r_2, p_2)$

(b) State of system if p_2 assigned r_2: cycle.

Figure 4-9. Resource allocation digraph for deadlock prevention.

4-5-3 Pert (Longest Paths)

Let $\{T_1, \ldots, T_n\}$ be a system of tasks where each task T_i takes time t_i, and the tasks are constrained by precedence constraints of the following type: task T_i can be initiated only after some specified subset of tasks $\{T_{i1}, \ldots, T_{im}\}$ has been completed. We can model such a system by a digraph $G(V, E)$ whose edges correspond to the tasks. The digraph is constructed so none of the tasks corresponding to edges emanating from a vertex v can be initiated until every task corresponding to an edge ending at v has been completed. The vertices of the digraph model themselves merely serve as loci where the edges that represent the precedence constraints can be brought together. If the precedence constraints are consistent, then the digraph model is acyclic. Refer to Figure 4-10 for an example. We are interested in the following kinds of questions:

1. What is the shortest time in which the system of tasks can be completed?
2. Is there any leeway in scheduling a task: What is the earliest time it can be started? What is the latest time it can be started without delaying the system?

The earliest time a task (represented by an edge (u, v)) can be started, consistent with the precedence constraints, equals the length of a longest path to u. While it is computationally infeasible to compute longest paths in general, there is a simple and efficient procedure, based on topological order, available in the case of acyclic digraphs. For simplicity, we will assume that the vertices are already topologically ordered and that there is a vertex v_1 from which all vertices are reachable and a vertex v_n reachable from every vertex, where n equals the order of G. We compute the lengths $Long(v, v_n)$ of the longest paths from each vertex v to v_n. The data definitions are suggestive only.

Procedure Longest (G)

(* Computes the lengths Long(i,n) of the longest paths from
each vertex v_i in G to vertex v_n *)

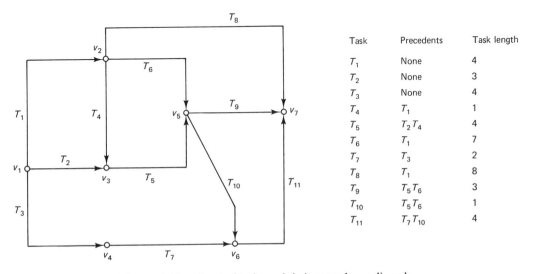

Task	Precedents	Task length
T_1	None	4
T_2	None	3
T_3	None	4
T_4	T_1	1
T_5	$T_2 T_4$	4
T_6	T_1	7
T_7	T_3	2
T_8	T_1	8
T_9	$T_5 T_6$	3
T_{10}	$T_5 T_6$	1
T_{11}	$T_7 T_{10}$	4

Figure 4-10. A set of tasks and their precedence digraph.

Trees and Acyclic Digraphs Chap. 4

```
var  G: Graph
     n: Integer constant
     k,j, Long(n,n): Integer

Set Long(n,n) to 0

for k = n − 1 to 1  do
    Set Long (k, n) to    max    { Len(k, j) + Long(j, n) }
                       (k,j) ∈ E(G)
End_Procedure_Longest
```

The lengths of the longest paths from v_1 to v can be computed similarly. We can define a variety of scheduling parameters in terms of the parameters Long, such as *Early(v)*, the earliest time at which a task starting at v can start, and *Late (v)*, the latest time at which a task starting at v can start without increasing the earliest possible completion time of the overall system of tasks. In terms of *Long*,

$$(1)\ Early(v) = Long(v_1, v)$$

$$(2)\ Late(v) = Long(v_1, v_n) - Long(v, v_n)$$

Tasks for which *Early* and *Late* are equal have no leeway in their scheduling. When such a task is delayed, the completion time of the whole system of tasks is increased. For this reason, these tasks are called *critical tasks,* and the paths along which they lie, which are precisely longest paths from v_1 to v_n, are called *critical paths*. Refer to Figure 4-11 for an example.

4-5-4 Optimal Register Allocation (Tree Labeling)

A classic problem of compilation is the so-called *register allocation problem:* determine the minimum number of registers required to evaluate an algebraic expression under the assumption that no stores into memory of intermediate operands are allowed. We can analyze this problem using a digraph model of expressions.

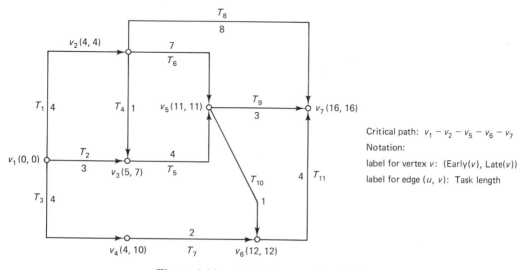

Critical path: $v_1 - v_2 - v_5 - v_6 - v_7$

Notation:

label for vertex v: (Early(v), Late(v))

label for edge (u, v): Task length

Figure 4-11. Schedule parameters for G.

Given an algebraic expression consisting of binary operators and operands, we can model the expression by a digraph where each operator and each distinct operand of the expression is represented by a distinct vertex of the digraph and there is an edge (x, y) between the vertices for an operator x and an operand y if x operates on y in the expression. Refer to Figure 4-12 for an example. Observe that the vertices of out-degree two correspond to binary operators, while vertices of out-degree zero correspond to operands. Operator vertices can be thought of as combining the values at their children. Operands such as x in Figure 4-12, which are the children of more than one vertex, are operated on by each of their parent operators. The induced subdigraph of vertices reachable from an arbitrary vertex v corresponds to the computation of a subexpression. We can think of the result of the subcomputation as available at v upon completion, so v behaves like an intermediate operand of the total computation. The model digraph may not be a tree, because of multiply occurring operands, but is acyclic.

A well-known graphical pebble game on the expression digraph models the register allocation problem. The rules of the game and its computational interpretation follow. We assume there is an unlimited supply of pebbles. The pebbles correspond to registers available for computing the expression represented by the digraph. Moving pebbles on the digraph corresponds to allocating and deallocating machine registers required by the computation. The pebble game moves are as follows.

(1) Place an available pebble on an endpoint.
(2) If there are pebbles on every child of v, then place one of the pebbles on v.
(3) If there are pebbles on every child of v, place a new pebble on v.
(4) Remove a pebble from a vertex v, making it available.

The pebble movements can be interpreted in terms of the machine instructions required for the evaluation of the expression.

(1) Load a register with a value.
(2) Take a binary function of the values in a pair of sibling registers and store the result in one of the registers, if the register is available at this point.
(3) Take a binary function of the values in a pair of sibling registers and store the result in a newly allocated register, if none of the previously allocated registers are available at this point.
(4) Make a register available for reuse.

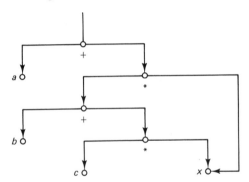

Figure 4-12. Acyclic digraph model for $a + x * (b + c * x)$.

Trees and Acyclic Digraphs   Chap. 4

The object of the pebble game is to compute the digraph by making a sequence of moves that, starting with an empty digraph, places a pebble on every vertex exactly once, and ends up with a vertex of in-degree zero pebbled. An example is shown in Figures 4-13 and 4-14.

It can be shown that the problem of determining the minimum number of pebbles to compute a digraph is NP-Complete. (Refer to Chapter 10 for this terminology.) Consequently, though the pebble game can be solved by enumeration for sufficiently small digraphs, both it and the register allocation problem it models are computationally intractable for large digraphs.

However, when the model digraph is a tree, there is a simple, efficient algorithm for the register allocation problem. Thus, following the notation of Figure 4-15, let $M(R)$ denote the minimum number of registers required to compute an expression represented by a binary tree model with root R without (!) stores. We can compute M recursively by the following labelling algorithm.

if M(Left(R)) = M(Right(R))

then Set M(R) to M(Left(R)) + 1

else Set M(R) to max { M(Left(R)), M(Right(R)) }

We set M to 1 for a left endpoint and to 0 for a right endpoint. Refer to Figure 4-16 for an example.

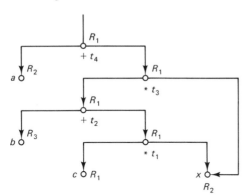

Figure 4-13. Digraph model with registers indicated.

Instruction	Type of Move	Vertex Pebble Placed On
L R1, c	1	c
L R2, x	1	x
M R1, R2	2	t1
L R3, b	1	b
A R1, R3	2	t2
M R1, R2	2	t3
L R2, a	4,1	a
A R1, R2	2	t4

Figure 4-14. Trace of pebble game for expression of Figure 4-12.

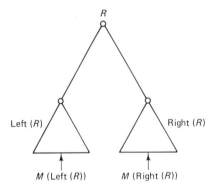

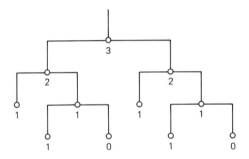

Figure 4-15. Recursive structure of register requirements algorithm.

Theorem (Correctness of Labelling Algorithm). Let T be a rooted binary tree (with root R) representing an expression. Then, the labelling algorithm correctly calculates the minimum number of registers $M(R)$ needed to evaluate the expression without using stores to memory.

The proof is as follows. Let T with root R be the smallest binary tree for which the theorem fails. That is, suppose T is the smallest tree that can be evaluated using fewer registers than $M(R)$. Let us denote the minimum number of registers required by T by m, and the minimum number of registers required by Left(R) and Right(R) by m_1 and m_2, respectively. We distinguish three possibilities:

$$(1) \; m_1 = m, \qquad m_2 < m,$$

$$(2) \; m_1 < m, \qquad m_2 = m, \quad \text{and}$$

$$(3) \; m_1 = m - 1, \qquad m_2 = m - 1.$$

(Any other combinations can be easily shown to force m to be less than $m - 1$, a contradiction.) To prove the procedure correctly calculates m in case (1), observe that since the subtrees Left(R) and Right(R) are smaller than T, then m_1 and m_2 must both equal the values $M(\text{Left}(R))$ and $M(\text{Right}(R))$ found by the algorithm. Furthermore, since the computation of T entails the computation of Left(R), then m must be at least as great as $M(\text{Left}(R))$. But, this is precisely the value assigned by the algorithm to $M(R)$ in case (1). Therefore, m must equal $M(R)$, as was to be shown. The proof is similar in case (2). To prove the result in case (3), we argue as follows. Suppose the first instruction of the computation that computes T loads (some register) with an operand from, say, Left(R). Then, there must be at least one register retained for the computation of Left(R) up to the point where the root R of T itself is evaluated. Therefore, during the remainder of the computation, there are at most $m - 2$ registers avail-

Figure 4-16. Register requirements tree.

able for computing Right(R), contradicting the assumption that Right(R) requires at least $m - 1$ registers for its computation. This completes the proof of the theorem.

Applications such as those just described illustrate how graphs can be used to model, in an abstract and concise manner, the essential structural features of a problem which may not be obviously graphical in nature. This allows us to reformulate questions about the original problem as graph-theoretic questions about the graph model, which can then by attacked by the methods and procedures of algorithmic graph theory.

4-6 FIBONACCI HEAPS AND MINIMUM SPANNING TREES

We described how to use Fibonacci heaps in Chapter 3 to implement a fast version of Dijkstra's shortest path algorithm. Fibonacci heaps can also be used to implement a fast version of Prim's minimum spanning tree algorithm (see Section 4-4), leading to an algorithm of performance $O(|V| \log |V| + |E|)$ for a weighted graph $G(V, E)$. An even faster, but more complicated algorithm, due to Fredman and Tarjan and also based on Fibonacci heaps yields a minimum spanning tree algorithm of performance $O(|E|(1 + \min\{i \mid \log^{(i)} (|V|) \le 2|E|/|V|\}))$, where $\log^{(i)}$ is defined inductively by: $\log^{(0)} (n) = n$, and $\log^{(i+1)} (n) = \log(\log^{(i)} (n))$. We shall describe both algorithms.

Fibonacci heap version of Prim MST algorithm. The structure of Prim's algorithm is exactly like that of Dijkstra's algorithm. The idea is to repeatedly extend a partial minimum spanning tree T one vertex at a time, always selecting as the next vertex to add to T that nontree vertex v whose connecting edge (v, t) to T has least weight. That is, (t, v) is the least weight edge between T and $G - V(T)$. The extended tree includes both the vertex v and the connecting edge (t, v).

At any point in the operation of the algorithm, each nontree vertex w which is a neighbor of some vertex in T is assigned a cost $c(w)$, equal to the weight of the least cost edge from w to T. The endpoint of this edge in T is denoted by $m(w)$. The set of vertices neighboring T are maintained in a heap, which is heap ordered on the costs of the vertices. The minimum element in the heap is then the nearest nontree vertex to the tree.

The detailed operation of the algorithm is as follows. The costs c of the vertices are initialized to $+M$ (which plays the role of plus infinity). An arbitrary vertex v_0 is selected as the vertex from which the spanning tree will be grown, and is inserted in the heap with a cost of 0. We then iteratively delete the minimum vertex v from the heap, set its cost to $-M$, add it to the minimum spanning tree under construction, add its associated edge to the minimum spanning tree (except in the case of the startup vertex v_0 which has no associated edge), and then scan the neighbors of v as follows. The neighboring vertices fall into three categories: vertices not previously reached, vertices already reached and deleted from the heap, and vertices reached but not yet deleted from the heap. The vertices not yet reached are just those with cost equal to $+M$. If w is such a vertex, we set the cost of w to the weight of the edge (w, v) connecting w to the tree and insert w in the heap. It is now a recognized neighbor of the spanning tree. If the cost of w is $-M$, then w and its associated edge are already in the minimum spanning tree, so we ignore it. Otherwise, w has already been reached by the search but is not yet in the minimum spanning tree; w has some current associated edge (w, u), where

u equals $m(w)$. If $c(w, v) < c(w, u)$, we change the associated edge of w to (w, v) and decrease the cost of w to $c(w, v)$, using a heap decrease-key operation.

Observe the strict similarity between the operation of this algorithm and the operation of Dijkstra's algorithm. There is a distinguished tree (or subtree), corresponding in this case to the partial minimum spanning tree. There is an extended tree that includes not just the partial minimum spanning tree, but also all the vertices that have been reached by the search process. The vertices in this search tree which are not yet in the partial minimum spanning tree lie in the heap. Finally, there is an unknown set of vertices that have not yet been reached by the search process. The operations, such as decrease-key, are the same as in Dijkstra's algorithm. Thus the Fibonacci heap implementation is $O(|V| \log |V| + |E|)$.

Fredman-Tarjan Fibonacci heap MST algorithm. The Fredman-Tarjan MST algorithm is a variation on the basic Prim algorithm wherein we constrain the tree growing process by restricting the size of the vertex heap. Once the heap reaches a certain limit, say after it contains k vertices, we empty the heap and start constructing a new tree at a previously unexamined vertex. Each application of this process results in a partial spanning tree. The process is iterated until every vertex in the graph is included in some component of a spanning forest. Then, we condense each of the partial spanning trees in the spanning forest into a single (super) vertex, one per component of the forest. We then repeat the whole process on this new graph of supervertices. The algorithm terminates when some phase produces a single spanning tree.

We now describe this process in more detail. Each vertex is assigned a cost, which is initially $+M$, and a mark field, which is initially set to unmarked. We construct a tree by selecting an unmarked vertex v_0, set cost(v_0) to 0, and insert v_0 into the heap. We then grow the tree starting at v_0 by repeating the following steps until either the size of the heap exceeds k or the heap is emptied or the tree being grown is linked to a previously grown tree.

(1) Delete the smallest element v from the heap and set cost(v) to $-M$; otherwise go to (5) if the heap is empty.

(2) If v is not equal to v_0, add the associated edge $(v, m(v))$ to the tree.

(3) If v is marked, and so the tree being grown has just been linked by $(v, m(v))$ to a previous tree, go to (5).

(4) Otherwise, mark v and scan each neighbor w of v as follows.
   (a) If cost$(w) = -M$, then w is already in the current tree and so, to prevent self-loops, we ignore w.
   (b) If cost$(w) = +M$, then w has not been scanned before; so we set cost(w) to $c(v, w)$, set $m(w)$ to v, and insert w in the heap.
   (c) If cost$(w) <> +M$ or $-M$, then w is already in the fringe of vertices neighboring the current tree. If cost$(w) > c(v, w)$, we decrease the cost of w to $c(v, w)$ and set $m(w)$ to v.
   At this point, if another unscanned neighbor of v remains and the heap size is less than k, we repeat steps (a) through (c) for that neighbor; otherwise, if an unscanned neighbor remains and the heap size equals k, go to (5); otherwise go to (1).

(5) Empty the heap (if necessary) and reset the keys of all the vertices scanned during the current phase to $+M$.

This tree-building process (selecting an unmarked vertex and repeating steps (1) to (5)) is repeated until every vertex is included in some tree (that is, is marked). When no unmarked vertex remains, we have constructed a spanning forest of the graph. Then, we condense the graph by shrinking each tree to a super-vertex. We then repeat the whole process on this new graph of super-vertices, resulting in a new spanning forest for this condensed graph. The process terminates only when some phase produces a trivial (single vertex) graph.

Condensing a graph $G(V, E)$ for which we have a current spanning forest can be done in time $O(|E|)$. Assume that for each vertex v, we know the number num(v) of the tree containing v. We can then lexicographically sort the edges on the tree numbers of their endpoints using a two-pass radix sort. Thus, we construct an array of $|V|$ queues, which are initially empty, and scan the second coordinates of the edge list, enqueueing each edge (u, v) on the queue for num(v). Then, we concatenate the $|V|$ queues in the array of queues to form a new edge list. This takes time $O(|V| + |E|)$. We then reinitialize the array of queues and scan the edge list on the first coordinate, inserting (u, v) on the queue for num(u). Repeating the concatenation process leaves the edges lexicographically sorted on the numbers of the trees of their endpoints. At this point, it is trivial to eliminate edges both of whose vertices are in the same tree, as well as to delete all but the least weight edge between a pair of distinct trees. The whole process takes $O(|V| + |E|)$ time.

Theorem (Fredman-Tarjan MST Performance). Let $G(V, E)$ be a weighted graph. Then, the Fredman-Tarjan minimum spanning tree algorithm takes $O(|E|(1 + \min\{i \mid \log^{(i)}(|V|) \le 2|E|/|V|\}))$ time.

The proof is as follows. We must first make a suitable choice for the heap size limit k. The choice involves a trade-off. The larger k is, the longer the passes will be, though there will be fewer of them. The smaller k is, the shorter the passes will be, but there will be more of them. If the current (condensed) graph contains t super-vertices, the algorithm will incur at most t **deletemin** operations and at most $O(|E|)$ inserts and decrease-key operations. If the heap size is at most k, the total time for the pass will be $O(t \log(k) + |E|)$. If we take k equal to $2^{(2|E|/t)}$, the running time of the pass is $O(|E|)$.

To bound the number of passes we argue as follows. Consider a pass where the number of trees equals t and the number of edges in the condensed graph is $|E'|$. Except for the final pass, the trees grown during a pass become blocked either because the heap size limit is reached or a pair of trees merge. In either case, the trees end up with at least k edges incident with the heap or fringe vertices. If a pass finishes with t' trees, there must be at least kt' such edges, which is at most $2|E'|$. Thus, $t' \le 2|E'|/k$. The heap size limit k' for the next pass then satisfies $k' = 2^{(2|E'|/t')}$ which is therefore at least 2^k. Considering the heap size limit for the initial pass, and since by the definition of the algorithm the size limit equals $|V|$ only on the final pass, it follows that the number of passes is bounded by $1 + \min\{i \mid \log^{(i)}(|V|) \le 2|E|/|V|\}$, as required. This completes the proof.

REFERENCES AND FURTHER READING

Deo (1974) has a thorough discussion of the matrices and vector spaces associated with graphs. Kruskal (1956), Prim (1957), and Dijkstra (1959) developed minimum spanning tree algorithms. A history of the problem is given in Graham and Hell (1985). For a discussion of path compression for unary trees, see Tarjan (1983), as well as Tarjan (1979), Banachowski (1980), and Tarjan and van Leeuwen (1984). Flajolet and Odlyzko (1982) obtain results on the average heights of random binary trees using generating function methods. Devroye (1986) gives the average height of a binary search tree. See also Knuth Vol. 3 (1973). The parts explosion application is from Gotlieb and Gotlieb (1978). See Coffman and Denning (1973) for a discussion of the optimal task scheduling algorithm, and further references, and McHugh (1984) for a short proof of the correctness of the algorithm. Sethi (1975) and Sethi and Ullman (1970) discuss the complete register allocation model. An interesting example of the use of fundamental cycles for determining the optimal placement of program probes is given in Knuth and Stevenson (1973). The minimum spanning tree algorithms based on Fibonacci heaps are from Fredman and Tarjan (1987). They improve earlier bounds, such as the $O(|E| \log(\log(|V|)))$ bound of Yao (1975), for sufficiently sparse graphs.

EXERCISES

1. Write a program that accepts a graph $G(V, E)$ and a spanning tree T of G as its input, and outputs the fundamental cycles of G with respect to T. Also, if a cycle C is additionally input, the program outputs the subset of fundamental cycles whose symmetric difference equals C.

2. Prove every cut is the union of edge disjoint minimal cuts. This is not true of disconnecting sets in general. Give an example.

3. Prove that in an eulerian graph a cutset must have an even number of edges.

4. Prove that for a connected graph $G(V, E)$
   (a) G is eulerian if and only if all its cuts are even, and
   (b) G is eulerian if and only if all its fundamental cuts are even.

5. Prove the number of endpoints in a nontrivial tree equals

$$1 + \left(\sum_i |d_i - 2| \right) / 2 ,$$

   where d_i's represent the degrees. Give an algorithm that counts the number of endpoints in a binary tree which is represented in a purely linked manner (like a binary search tree) so that its degree sequence is indeterminate. Make the algorithm also determine the height of the tree.

6. Characterize all structured flowcharts algorithmically by showing how they can be derived from an initial primitive structured flowchart using suitable vertex and edge insertions.

7. Graph-theoretically characterize the minimum number of counters that need to be maintained to monitor the frequency of execution of the statements in a structured program. Use the linear dependence of the values of the counters. Where do the optimal counters have to be placed?

8. Try to find an efficient algorithm for determining whether reversing a given edge in an acyclic digraph introduces a cycle.

9. Write an algorithm that generates random permutations, and then use these to generate random binary search trees. Compare the average height of these trees with Devroye's estimate.

10. Prove that the euclidean minimum spanning tree constructed using the Voronoi diagram has maximum degree equal to 5.

11. Give an interpretation for the minimum spanning tree problem, where the edge weights are

probabilities of edge failure and the objective function is the product of the edge weights of the spanning tree. Does the same algorithm work?

12. Consider the following greedy algorithm for euclidean spanning trees in 2-space. First, sort the vertices on their x coordinates, then recursively construct a minimum spanning tree on each half of the vertices; then join the two subtrees by the shortest edge connecting them. Comment on correctness, error estimate, counterexample, and performance.

13. Prove that if d_i, $i = 1, \ldots, n$ is a sequence of positive integers whose sum equals $2n - 2$, there exists a tree with this sequence as its degree sequence.

14. Prove the complement of a tree is either connected or consists of an isolated vertex and a complete subgraph.

15. What is the long-term effect of an error in the transmission of a bit in a Huffman encoded message?

16. Implement the minimum spanning tree algorithm using Voronoi diagrams.

17. Let $G(V, E)$ be a graph. If $\min(G)$ is at least k, then G contains every tree on $k + 1$ vertices as a subgraph.

18. Write an efficient implementation of the algorithms for the PERT parameters described in the text.

19. Compare the performance of implementations of the recursive and nonrecursive algorithms given for topological sort.

20. Why does the longest path algorithm break down if the digraph is not acyclic?

21. Show the minimum number of comparisons needed to insert a number into an ordered list of n numbers is $O(\log n)$.

22. What is the maximum number of edges in an acyclic digraph?

23. Let $G(V, E)$ be a graph satisfying that every pair of its vertices have a unique neighbor in common. Characterize G (first show it has a spanning star).

5

Depth-First Search

5-1 INTRODUCTION

We traverse a data structure by visiting all its nodes and pointers. For example, there are several standard techniques for traversing a binary tree. If the tree is a binary search tree, we use inorder traversal to list its search keys in order. If the tree represents an expression in infix form, we use preorder traversal to convert the expression to prefix form.

Graph traversal techniques vastly generalize these ideas. We consider arbitrary graphs instead of ordered binary trees, and we typically assume no special ordering for the edges incident at a vertex of the graph. Graph traversal procedures show how to systematically explore a graph, both to access information stored at its vertices and edges, and to identify structural properties such as whether or not the graph is connected.

Depending on how a graph is represented, it may or may not be easy to even identify its vertices and edges. For example, if a graph is represented as a linear array, we can directly access its vertices and sequentially access its edge lists, although even so simple a property as connectedness will not be obvious from the representation. On the other hand, if a graph is represented by a pure linked representation, where we must follow pointers to access the graph, it requires care even to find all the vertices and edges in the graph.

The two most common techniques for traversing graphs are depth-first search and breadth-first search. In *depth-first search,* we repeatedly extend a search path as far as possible into a graph, retract it, and then reextend it in another direction, until we have traversed the whole graph. In *breadth-first search,* on the other hand, we explore the graph in a level by level fashion, starting the search at a given vertex, and progressively extending the search to vertices at greater distances.

Depth-first search is useful in identifying structural properties of graphs. On the other hand, for problems such as maximum flow (which we will consider in Chapter 6), where the structural characteristics of depth-first search are not needed, breadth-first search may lead to better performance.

Despite their different graphical interpretations, the implementations of the algorithms for depth- and breadth-first searches are quite similar, differing only in the data structure used to control the search. Depth-first search uses a stack for this purpose, while breadth-first search uses a queue. As a consequence, depth-first search can be implemented naturally using recursion. Our discussion will concentrate on depth-first search and some of its variations.

5-2 DEPTH-FIRST SEARCH ALGORITHMS

We will initially describe depth-first traversal in its simplest form, where we number vertices according to the order in which they are visited without classifying edges. We will then augment the algorithm to classify edges. For undirected graphs, the edge classification is simple. For directed graphs, the classification is more complex. In subsequent sections, we will show how the edge classifications can be used to analyze the structure of graphs.

5-2-1 Vertex Numbering

We will first informally describe depth-first search. The procedure is the same for both undirected and directed graphs.

(1) We start the search at a vertex v and initialize the search path to v.

(2) If there is an unexplored edge at the vertex at the head of the search path leading to an unexplored vertex w, we extend the search path to w, which becomes its new head, and repeat step (2).

(3) If there is an unexplored edge at the head of the search path leading to an explored vertex w, we mark that edge as explored and repeat (2) for the next edge incident at the head of the search path.

(4) If there are no further unexplored edges at the head of the search path, we retract the search path to the previous vertex by removing the current head of the path and repeat step (2).

We stop once the search path is empty.

The name "depth-first search" derives from the pattern of the search path which always advances from its last explored vertex, thus making the path drive deeper into the graph before it explores closer regions of the graph.

We will now describe a formal procedure for depth-first search. We represent the graph or digraph as a linear array $\mathbf{H}(|\mathbf{V}|)$ with components of type **Vertex**.

```
type   Vertex = record
                    Dfsnum: 0..|V|
                    Positional Pointer: Edge pointer
                    Successor: Edge pointer
                end
       Edge   = record
                    Neighbor: 1..|V|
                    Successor: Edge pointer
                end
```

Dfsnum holds the depth-first search number of the vertex and is initially zero. As usual, the positional pointer for x points to the last explored edge at x. We call the graph (digraph) data type Graph. We use a stack S to represent the search path. Its entries have the following form.

```
type   Stack Entry = record
                        Index: 1..|V|
                        Successor: Stack Entry pointer
                     end
```

The depth-first search procedure calls a variety of functions and other procedures. We use standard *stack operations*.

(S1) Create(S)

(S2) Push(S,v)

(S3) Top(S)

(S4) Pop(S)

(S5) Empty(S)

Create(S) sets S to a nil stack entry pointer. Push(S,v) creates a stack entry for v and pushes it on the stack S. Top(S) returns in Top the index of the next vertex on the stack. Pop(S) deletes the top of the stack S and also returns a pointer to the new top in Pop. Empty(S) succeeds when the stack is empty, and fails otherwise.

The function Next(x,y) returns in y the index of the next neighbor of vertex x and fails if there is none. Nextcount is an auto-increment integer function which is initially zero.

The control logic of the procedure mirrors the advance and retraction of the search path. The **while** loop advances the search path until it is blocked at a vertex all of whose neighbors have been explored by the time we reach the vertex. The outer loop retracts the search path by backing it up to the previous vertex on the search path stack. The search terminates when the search path is empty. Figure 5-1 gives a flowchart for the algorithm.

```
Procedure DFS(G,u)

(* Performs DFS traversal of vertices in G reachable from u,
     where G is either an undirected or directed graph *)


var   G: Graph
      S: Stack Entry pointer
      u, v: 1..|V|
      Nextcount, Top: Integer function
      Next, Empty: Boolean function
      Pop: Stack Entry pointer function


Create (S)
Set Dfsnum(u) to Nextcount
Push (S, u)


repeat


   while   Next (Top(S), v) do
```

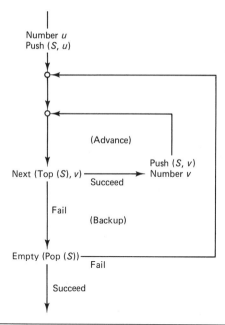

Figure 5-1. Depth-first search using explicit stack.

> **if** Dfsnum(v) = 0 **then** (* Advance Search Path *)
> **Set** Dfsnum (v) to Nextcount
> Push (S, v)

> **until** Empty(Pop(S)) (* Retract Search Path or Exit *)

End_Procedure_DFS

Recursive depth-first search.   We can also specify the depth-first search procedure recursively. This simplifies the procedure because the search path stack is maintained automatically and implicitly as part of the recursion. Each activation of the recursive procedure corresponds to an advance of the search path, while each return from a recursive activation corresponds to a retraction of the search path. This makes the control logic for the backtracking features of the algorithm transparent. The recursive DFS procedure is as follows.

> **Procedure** DFS(G,u)
>
> (* Performs recursive depth-first search from u *)
>
> **var**   G: Graph
> u, v: 1..|V|
> Nextcount: Integer function
> Next: Boolean function
>
> **Set** Dfsnum(u) to Nextcount
>
> **while**   Next(u,v) **do** **if**   Dfsnum(v) = 0 **then**   DFS(G,v)
>
> **End_Procedure_**DFS

The argument u may be either a value (that is, input only) parameter, or a reference (that is, input and output) parameter. The variable v in the procedure should be declared as a local variable. Then, each time DFS invokes itself, a new instance of v is allocated, with its own storage, local to that activation. The search path stack is implicitly stored by the recursion mechanism as the sequence of such allocations and so is transparent. If we make u a value parameter, recursion also provides a stack in which the successive values of u are saved, corresponding to a second (redundant) copy of the search path stack.

Breadth-first search.　If we change the data structure used to store vertices from a stack to a queue, we obtain breadth-first-search. A breadth-first-search visits vertices in a level-by-level fashion, nearer vertices first. For example, if the graph is a tree, breadth-first-search visits the root (at level zero) first, then the vertices at level one (in some order), and so on. For an arbitrary graph, the vertices at level i of the search are precisely the vertices at distance i from the initial vertex.

The algorithm uses the standard queue operations.

(Q1) Create(Q)

(Q2) Enqueue(Q,v)

(Q3) Head(Q)

(Q4) Dequeue(Q)

(Q5) Empty(Q)

Their definitions are analogous to the corresponding stack operations. The algorithm is as follows.

```
Procedure BFS(G,u)

(* Breadth-first traversal of vertices in G reachable from u *)

var  G: Graph
     u, v: 1..|V|
     Nextcount: Integer function
     Next, Empty: Boolean function
     Q: Queue entry pointer
     Head, Dequeue: Queue entry pointer function

Create(Q)
Set Bfsnum(u) to Nextcount
Enqueue(Q,u)

repeat

     (* Number every unvisited vertex adjacent to Head(Q) *)

     while  Next(Head(Q),v)  do

         if Bfsnum(v) = 0 then   Set Bfsnum(v) to Nextcount
                                 Enqueue(Q,v)
```

until Empty(Dequeue(Q))   (* Retreat or Exit *)

End_Procedure_BFS

A breadth-first search numbering of the vertices of the graph in Figure 5-2a is shown in Figure 5-2c. The depth-first labels are shown in Figure 5-2b. Refer to the exercises for a procedure that determines the breadth-first search level of a vertex as well.

Induced depth-first search tree.   The depth-first traversal from a vertex u in a graph (or digraph) G implicitly determines a directed tree T rooted at u which we call the *depth-first search tree*. $V(T)$ consists of every vertex reachable from u. An edge (x, y) is in $E(T)$ if y was as yet unexplored when (x, y) was explored from x. We consider (x, y) as directed from x to y, making T a directed tree rooted at u. This corresponds to the built-in direction of the edge when G is directed, and to an induced direction when G is undirected. We can easily augment any of the procedures for per-

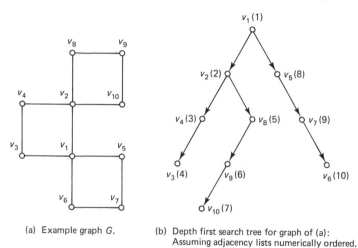

(a) Example graph G.

(b) Depth first search tree for graph of (a): Assuming adjacency lists numerically ordered.

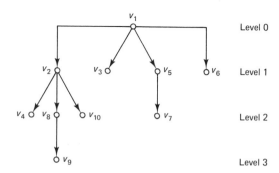

(c) Breadth first search tree for (a): Assuming adjacency lists numerically ordered.

Figure 5-2.   Depth- and breadth-first search examples.

forming a depth-first search so they also actually construct the depth-first search tree. The critical structural feature of depth-first search is given by the following theorem.

Theorem (Depth-First Search Structure Theorem). Let $G(V, E)$ be a graph (digraph) on which a depth-first traversal is performed, creating a directed tree $T(u)$ rooted at an initial vertex u. Let x and y be any pair of vertices reachable from u. Let $T(x)$ and $T(y)$ denote the subtrees of $T(u)$ rooted at x and y respectively. If Dfsnum(x) < Dfsnum(y),

(1) Either y is in $T(x)$, or

(2) $T(x)$ and $T(y)$ are disjoint, and there is no edge in G from $T(x)$ to $T(y)$.

It is this feature that makes depth-first searching useful in identifying the structure of graphs, especially those properties related to connectivity. The breadth-first search satisfies no similar structural property, though we can still define a *breadth-first search induced tree* analogous to the depth-first search induced tree.

5-2-2 Edge Classification: Undirected Graphs

The applications of depth-first search depend on the classification of edges that it leads to. The classifications differ depending on whether the graph is directed or not. We will consider undirected graphs first.

We will classify the edges of undirected graphs into two types: tree edges and back edges. The *tree edges* are the edges in the induced depth-first search tree. Recall that we said an edge (x, y) was an edge of the induced tree if y was unexplored when we explored (x, y) from x. For consistency, (x, y) ought to be classified the same whether viewed from x or from y. By definition, a *back edge* is any edge which is not a tree edge. Once again, consistency demands that both representatives of a back edge be classified identically.

Since G is undirected, we must take care that redundantly represented undirected edges are classified uniquely. That is, since the distinct representatives of an edge refer to the same undirected edge, they ought to be classified identically in any consistent classification scheme.

The traversal examines each undirected edge twice, once at each representative of the edge. We can distinguish between tree and back edges by distinguishing between the first and second explorations of an undirected edge (x, y).

(1) Tree edge, First visit

(2) Tree edge, Second visit

(3) Back edge, First visit

(4) Back edge, Second visit

Figure 5-3 shows a scheme for classifying an edge (x, y).

Back edges satisfy the following important property.

Theorem (Back Edge Property for Undirected Graphs). Let $G(V, E)$ be an undirected graph, and let T be an induced depth-first search tree on G. Let (u, v) be a back edge with respect to T. Then:

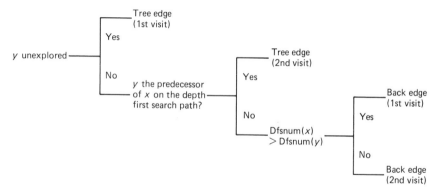

Figure 5-3. Edge classification decision tree for edge (x, y).

(1) The vertices u and v stand in an ancestral/descendant relationship in T.

(2) If Dfsnum(u) < Dfsnum(v), u is an ancestor of v; otherwise v is an ancestor of u.

(3) The edge (u, v) is first explored from whichever of the vertices is the descendant vertex.

We can use the edge classification to induce a direction on the edges of G as follows:

(1) The tree edges are directed from tree parent to tree child;

(2) The back edges are directed from descendant to ancestor, or equivalently, from the vertex with the higher Dfsnum to the vertex with the lower Dfsnum.

This defines an *induced digraph* which we will consider further when we examine the problem of graph orientability.

The following algorithm classifies the edges of an undirected graph G. We modify our previous Edge data type to

```
type   Edge   = record
                  Neighbor: 1..|V|
                  Edge Type: (Tree, Back)
                  Successor: Edge pointer
               end
```

A function Predecessor_of_Top returns the value of Top(Pop(S)), or zero if there are less than two elements on the stack.

```
Procedure DFS_EC(G,u)

(* Depth-first search edge classification: undirected graphs *)

var   G: Graph
      S: Stack Entry pointer
      u, v: 1..|V|
      Nextcount, Top, Predecessor_of_Top: Integer function
      Next, Empty: Boolean function
      Pop: Stack Entry pointer function
```

```
Create(S)
Set Dfsnum(u) to Nextcount
Push (S,u)


repeat

    while Next(Top(S),v) do

        if   Dfsnum(v) = 0

        then   (* Tree edge, 1st visit *)
               Set Edge Type (Top(S),v) to Tree
               Set Dfsnum(v) to Nextcount
               Push(S,v)

        else   if   v = Predecessor_of_Top(S)

               then   (* Tree edge, 2nd visit *)
                      Set Edge Type (Top(S),v) to Tree

               else   (* Back edge, either visit *)
                      Set Edge Type (Top(S),v) to Back


    until Empty(Pop(S))

End_Procedure_DFS_EC
```

We will now present a recursive version of the edge classification algorithm. The algorithm, **DFS_EC**, uses an explicit stack S to differentiate between a tree edge on its second visit and a back edge. This explicit stack was unnecessary in the pure vertex numbering algorithm because the distinction was not required then. We will assume S is initially empty.

```
Procedure DFS_EC(G,S,u)

(* Recursive DFS edge classification from u: undirected graph *)

var   G: Graph
      S: Stack Entry pointer
      u, v: 1..|V|
      Nextcount, Top, Predecessor_of_Top: Integer function
      Next: Boolean function

Set Dfsnum(u) to Nextcount
Push (S,u)

while Next(u,v) do

    if     Dfsnum(v) = 0

    then   (* Tree edge, 1st visit *)
           Set Edge Type (u,v) to Tree
           DFS_EC(G,S,v)
```

```
              else    if v = Predecessor_of_Top(S)
                      then   (* Tree edge, 2nd visit *)
                             Set Edge Type (u,v) to Tree
                      else   (* Back edge, either visit *)
                             Set Edge Type (u,v) to Back

       Pop(S)
       End_Procedure_DFS_EC
```

The explicit stack is used only to identify the predecessor of Top(S). We can avoid the stack by retaining the top two vertices of the search path at each invocation. The implicit recursion stack will still maintain the complete search path, which is required for backtracking. The revised procedure DFS_EC(G,w,u) follows. The procedure is initially invoked as DFS_EC(G,nil,u).

```
       Procedure DFS_EC(G,w,u)

       (* Recursive DFS from u with DFS search predecessor w *)

       var  G: Graph
            u, v, w: 1..|V|
            Nextcount, Top: Integer function
            Next: Boolean function

       Set Dfsnum(u) to Nextcount

       while Next(u,v) do

          if   Dfsnum (v) = 0

          then   (* Tree edge, 1st visit *)
                 Set Edge Type (u,v) to Tree
                 DFS_EC(G,u,v)

          else   if v = w

                 then   (* Tree edge, 2nd visit *)
                        Set Edge Type (u,v) to Tree

                 else   (* Back edge *)
                        Set Edge Type (u,v) to Back

       End_Procedure_DFS_EC
```

5-2-3 Edge Classification: Directed Graphs

The classification of edges for directed graphs is more diverse, but since the edges are represented nonredundantly, the classification is automatically consistent. We classify a directed edge (x, y) with respect to a given depth first traversal as

(1) A *tree edge* if y is unexplored when (x, y) is explored from x,

(2) A *back edge* if y is an ancestor of x with respect to the depth first tree,

(3) A *forward edge* if y is a descendant of x in the depth first tree, or

(4) A *cross edge* if neither x nor y is a descendant of one another.

Not every vertex may be reachable from a given root u. Therefore, several depth first searches, each starting from a different vertex, may be needed to explore the whole graph. Each search determines a corresponding (disjoint) depth-first tree. Except for cross edges, the edges are always between vertices of a single depth-first tree. A cross edge may connect either a pair of vertices within a single tree or a vertex in one search tree to a vertex in a previously traversed search tree. A procedure for classifying the edges is

(1) If y is unexplored, (x, y) is a tree edge, by definition;

(2) Otherwise, if Dfsnum(y) > Dfsnum(x), (x, y) must be a forward edge; since y cannot be an ancestor of x, (x, y) is not a back edge; and (x, y) cannot be a cross edge since Dfsnum(y) exceeds Dfsnum(x):

(3) Otherwise, (x, y) is a back or cross edge, depending on whether y is an ancestor of x or not; if y is still on the search path when explored from x, then y is an ancestor of x; so (x, y) is a back edge. Otherwise, (x, y) is a cross edge.

Figure 5-4 shows how to classify an edge (x, y) being explored from a vertex x, and Figure 5-5 gives an example. These results are summarized in the next theorem.

Theorem (Classification of Edges for Digraphs). Let $G(V, E)$ be a directed graph and suppose the edges of G have been classified with respect to some depth first search traversal. Let (x, y) be an edge of G. Then,

(1) Dfsnum(x) < Dfsnum(y) if and only if (x, y) is a tree or forward edge, and

(2) Dfsnum(x) > Dfsnum(y) if and only if (x, y) is a back or cross edge.

The edge classification can be used to determine structural properties of a graph, as illustrated by the following.

Theorem (Cycle Detection by Depth First Search). Let $G(V, E)$ be a directed graph, and suppose the edges of G have been classified with respect to a depth-first search traversal. Then, G is acyclic if and only if no edge of G has been classified as a back edge.

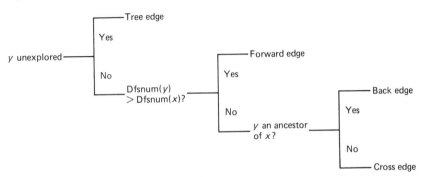

Figure 5-4. Edge classification for directed graphs.

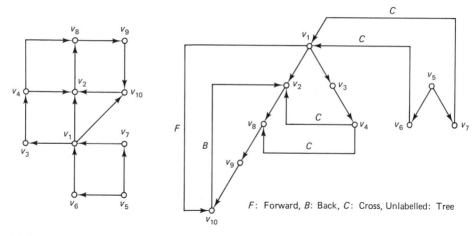

(a) Directed depth first search
    edge classification.

(b) Directed edge classification for graph of (a).

Figure 5-5. Directed edge classification.

The proof is as follows. Clearly, if there is a back edge, G is not acyclic. Suppose, on the other hand, that G contains a cycle with successive vertices $v_1, \ldots, v_n$. Let v_i be the first cycle vertex reached during the depth-first search. Assume for simplicity that $i > 1$. Then, v_{i-1} must be a descendant of v_i in the depth-first search tree. Therefore, the cycle edge (v_{i-1}, v_i) must be a back edge with respect to the depth-first search, as was to be shown.

The algorithm that classifies the edges follows. We will assume the Stack S is represented as an array, augmented by an indicator array **On**. **On**(v) is set to 1 if v is on the stack, and to 0 otherwise.

```
type   Stack = record
              SS(|V|): 1..|V|
              Top, On(|V|): 0..|V|
              end


Procedure DFS_EC(G,S,u)

(* DFS digraph edge classification with explicit stack *)

var    G: Graph
       u, v, w: 1..|V|
       S: Stack
       Nextcount: Integer function
       Next: Boolean function

Set Dfsnum(u) to Nextcount
Push(S,u)

while Next(u,v) do

       if      Dfsnum(v) = 0
```

```
        then   (* Tree edge *)
               Set Edge Type (u,v) to Tree
               DFS_EC(G,S,v)

    else   if Dfsnum(v) > Dfsnum(u)

        then   (* Forward edge *)
               Set Edge Type (u,v) to Forward

        else   if v is on search path stack S

                then   (* Back edge*)
                       Set Edge Type (u,v) to Back

                else   (* Cross edge *)
                       Set Edge Type (u,v) to Cross


    Pop(S)


    End_Procedure_DFS_EC
```

Alternative classification algorithms. The previous depth-first classification algorithm uses an explicit stack. There are alternatives. One is to use an array of switches, denoted **Completed** $(|V|)$, which is initially set to false for every vertex and is set to true for a vertex v once all the edges at v are explored, or equivalently, when the search path retracts from v. The switches can be used to differentiate between back and cross edges. We have:

Theorem (Switch Characterization of Cross and Back Edges). Let $G(V, E)$ be a directed graph. If an edge (x, y) of G is being explored during the execution of a depth-first traversal of G, then (x, y) is a cross edge if and only if

$$Dfsnum(y) < Dfsnum(x) \quad \text{and} \quad \textbf{Completed}(y) \text{ is true}.$$

(The test in this theorem must be applied during the execution of the depth-first search, since once the search is done, **Completed**(y) will be true for every vertex y. The associated traversal algorithm is straightforward.)

Another alternative is based on distinguishing between back and forward edges based on the order of completion of traversal of vertices. Recall that the variable *Dfsnum* ranks vertices according to the order in which they are added to the search path stack or first explored. We can define a complementary variable *Completion* which ranks vertices according to the order in which they are popped from the stack. Thus, while the first vertex added to the search path is assigned *Dfsnum* equal to 1, the first vertex the search path backs up from would be assigned *Completion* equal to 1. We assume that *Completion*(v) is initialized to some large integer M (which acts as plus infinity), for every vertex v. This ensures the *Completion* number of incompletely traversed vertices is greater than the *Completion* number of completely traversed vertices. We have the following theorem.

Theorem (Ancestor Characterization using Completion). Let $G(V, E)$ be a directed graph. Let (u, v) be an edge in G. Then,

(1) (u, v) is a back edge with respect to a given depth-first search if and only if $Dfsnum(v) < Dfsnum(u)$ and $Completion(u) \leq Completion(v)$.

(2) (u, v) is a cross edge with respect to a given depth-first search if and only if $Dfsnum(v) < Dfsnum(u)$ and $Completion(v) < Completion(u)$.

The depth-first search algorithm based on this theorem is straightforward.

5-3 ORIENTATION ALGORITHM

An undirected graph $G(V, E)$ is *orientable* if we can impose directions on its edges $E(G)$ (determining a new set of directed edges E') in such a way that the digraph $G'(V, E')$ is strongly connected. The digraph G' is called an *orientation* of the undirected graph G. Refer to Figure 5-6 for examples. If we think of G as a roadmap whose edges correspond to two-way roads, G is orientable if and only if there is some way in which we can make every road one-way and yet still be able to travel between any pair of points in G.

We will describe two algorithms for testing whether a graph is orientable. One approach uses depth first search, and is more efficient. The other is a straightforward application of the following graph-theoretic characterization of orientability. Recall that a *bridge* is an edge whose removal increases the number of components in a graph.

Theorem (Orientability Characterization). A connected graph $G(V, E)$ is orientable if and only if it contains no bridges.

This theorem implies that nonorientable connected graphs are precisely the graphs of edge connectivity one, and therefore an algorithm for testing orientability also tests whether $EC(G)$ is at most one. The necessity of the condition in the theorem is obvious. If the graph contains a bridge, it cannot be orientable because whatever direction

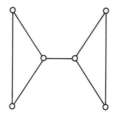

(a) Nonorientable graph.

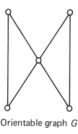

Orientable graph G Orientation of G

(b)

Figure 5-6. Orientation.

is assigned to the bridge will force access to be blocked in the opposite direction. On the other hand, it is not obvious that the absence of a bridge ensures orientability. This will follow from our proof of correctness of the depth-first search based orientation algorithm.

The exhaustive search algorithm suggested by the theorem is outlined in Figure 5-7. Its performance is determined by the frequency of execution of its search loop, which depends both on $|E(G)|$ and on whether the graph is orientable or not. If the graph is orientable, every edge must be examined before the algorithm terminates. If the graph is not orientable, the algorithm terminates as soon as a bridge is detected. Thus, in the worst case, the search loop is executed $O(|E|)$ times. The connectivity test embedded in the loop takes $O(|E|)$ time using depth-first search, hence the overall complexity of the algorithm is $O(|E|^2)$.

Depth-first orientation. We now describe an algorithm based on depth-first search that determines orientability in $O(|E|)$ steps. This algorithm also finds an orientation if one exists, in contrast to the exhaustive search algorithm which merely tests for orientability. The algorithm is based on the following idea. Any depth-first search traversal of an undirected (connected) graph G induces a directed graph G'. The tree edges of the traversal are directed in G' from lower to higher *Dfsnum*, and the back edges are directed from higher to lower *Dfsnum*. We can show that if G is orientable, then G' is an orientation of G; while if G is not orientable, the individual components of the graph G'', obtained from G by removing its bridges, are oriented.

We can decide which case has occurred by considering the induced depth first directed tree T rooted at r in G'. G' is an orientation of G if and only if

(1) There is a directed path in T from r to every vertex in G', and

(2) There is a directed path from every vertex in G' to r.

The first condition is automatically satisfied when G is connected, since T is a directed tree with root r. Therefore, we need only consider when (2) holds. We consider this question using a "trapping condition" we will define.

Let v be a vertex in G', and let $T(v)$ be the subtree of T rooted at v. We will say a vertex x in $T(v)$ is "trapped in $T(v)$" if there is no directed path in G' from x to any vertex outside $T(v)$. If not all vertices in $T(v)$ are trapped in $T(v)$, there must be a back edge from at least one vertex u in $T(v)$ to some vertex w not in $T(v)$. Since T is a depth-

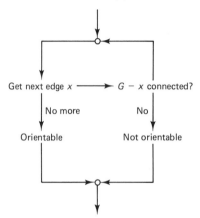

Figure 5-7. Exhaustive search algorithm for orientability.

first search tree, w must be an ancestor of v. Thus, if any vertex in $T(v)$ is not trapped in $T(v)$, there must be a directed path in G' from v to some vertex with a lower *Dfsnum* than v.

We will say a subtree $T(v)$ satisfies a nontrapping condition if none of its vertices are trapped in $T(v)$. If every subtree (except T itself, of course) satisfies a nontrapping condition, there is a directed walk from any vertex in G' to the root r of T. The walk is identified by merely repeatedly following the directed paths guaranteed to exist from each vertex to an ancestor vertex, until we eventually reach the root r, the vertex of least *Dfsnum*. Since, a directed walk between a pair of vertices necessarily contains a directed path between those vertices, then under these circumstances condition (2) holds. We refer to Figure 5-8 for an illustration.

The nontrapping condition motivates introducing, for every vertex v in G, a parameter *Low*(v) given by

$$\min\{\textit{Dfsnum}(w) \mid u \text{ is in } T(v) \quad \text{and} \quad (u, w) \text{ is a back edge}\}.$$

The interpretation of *Low*(v) is that it equals the *Dfsnum* of the earliest ancestor of v which is reachable from v by a directed path in $T(v)$ followed by a single back edge. In other words, it directly implements the nontrapping condition. We can compute *Low*(v) (refer to Figures 5-9 and 5-10) by combining a back edge rule which sets *Low*(v) to min $\{Low(v), Dfsnum(u)\}$, where the minimum is taken over all back edges (v, u) at v, and an inheritance rule which sets *Low*(v) to min $\{Low(v), Low(u)\}$, where the minimum is taken over all children u of v. The back edge rule has the effect of identifying the earliest ancestor reachable from a vertex in G' using only a single back edge. The inheritance rule includes in *Low*(v) the least *Dfsnum* reachable by any vertex in the subtree $T(v)$. Using *Low*, we can then formulate the nontrapping condition as

Nontrapping Condition: If *Low*(v) < *Dfsnum*(v) for every vertex v not equal to the root r, then G is orientable.

We can show conversely that if there is some subtree $T(v)$ (other than T itself) for which v is trapped in $T(v)$, then v is incident with a bridge, namely the entry edge to the subtree $T(v)$. The corresponding trapping condition follows. *Par*(v) denotes the parent vertex of v in the depth-first tree.

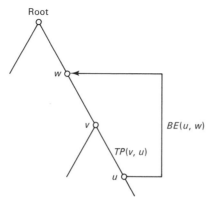

TP(v, u): Tree path from v to u
BE(u, w): Back edge from u to w

Figure 5-8. Induced path from v to some ancestor w of v.

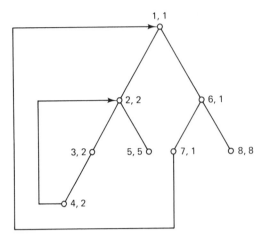

Figure 5-9. Labelling with (Dfsnum, Low).

Trapping Condition: If $Low(v) = Dfsnum(v)$ for some vertex v not equal to the root r, then $(par(v), v)$ is a bridge of G.

To prove this condition, suppose $Low(v) = Dfsnum(v)$ for some v not equal to r, and suppose that $(par(v), v)$ is not a bridge. Refer to Figure 5-11 for an illustration. Then, there must be a back edge (u, w) from some vertex u in $T(v)$ to an ancestor w of v. Consequently, $Low(v)$ must be at least as small as $Dfsnum(w)$, by the back edge and inheritance rules. Therefore, $Low(v)$ is less than $Dfsnum(v)$, contrary to assumption.

This completes the proof of the correctness of the conditions. We summarize the results in a theorem.

Theorem (Trapping/Nontrapping Condition For Orientability).   Let $G(V, E)$ be a connected graph, and let r be the root of a depth-first search tree in G. Then, if $Low(v) < Dfsnum(v)$ for every vertex v not equal to r, then G is orientable. On the other hand, if $Low(v) = Dfsnum(v)$ for some vertex v not equal to r, the edge $(par(v), v)$ is a bridge of G.

It is now easy to design a depth-first search procedure to compute Low and test the trapping condition. Refer to Figures 5-12 and 5-13 for an example.

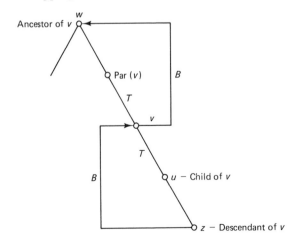

Figure 5-10.   Relations among ancestors, descendants, and children of v.

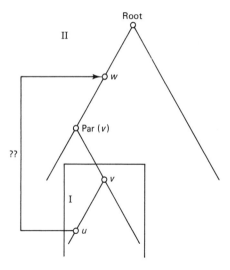

I: The subtree rooted at v
II: The remainder of the search tree

Figure 5-11. Illustration for trapping condition.

5-4 STRONG COMPONENTS ALGORITHM

We define a *strong component* of a digraph $G(V, E)$ as a connected induced subgraph of G of maximal order. The strong components of a digraph are analogous to the components of an undirected graph, corresponding to the maximal internally connected pieces of the digraph. See Figures 5-14 and 5-15 for an example. We present an algorithm for finding the strong components of a digraph based on depth-first search. The algorithm uses an enhanced depth-first search, similar to the orientation algorithm.

The depth-first organization of the algorithm is critical to its correct operation. It guarantees that whenever the search enters a region of the graph corresponding to a strong component and which has an exit edge to an adjacent strong component, the search continues into that adjacent component, and explores it fully, before returning to and completing the exploration of the earlier strong component. This behavior allows

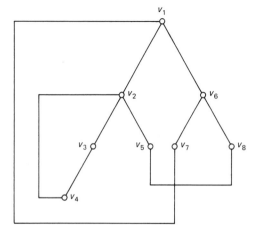

Figure 5-12. Graph for orientation example.

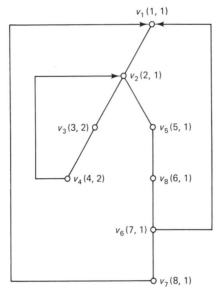

Figure 5-13. Graph with (Dfsnum,Low) values at each vertex.

us to traverse the strong components in a nested fashion, while identifying them sequentially.

Like the orientation algorithm, the strong components algorithm uses a parameter *Low* which aids in the recognition of the strong components in a manner similar to the recognition of bridges in the orientation algorithm. That is, the critical points occur where the search retracts from vertices v where $Dfsnum(v) = Low(v)$.

The basic data structures are a *search stack* SP and a so-called *component list stack* CL. SP is handled in the usual depth-first manner. The CL vertices correspond to vertices from partially traversed strong components. As the search progresses, it saves vertices on CL so that at any point CL contains every vertex reached since the last complete strong component was identified, as well as any previously reached vertices from strong components that have so far been only partially traversed. (Remember the depth-first traversal makes the search enter the strong components in a nested manner, so that

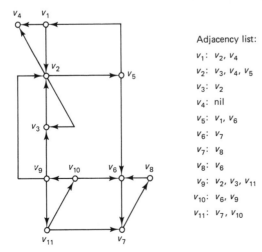

Adjacency list:

v_1: v_2, v_4
v_2: v_3, v_4, v_5
v_3: v_2
v_4: nil
v_5: v_1, v_6
v_6: v_7
v_7: v_8
v_8: v_6
v_9: v_2, v_3, v_{11}
v_{10}: v_6, v_9
v_{11}: v_7, v_{10}

Figure 5-14. Digraph G.

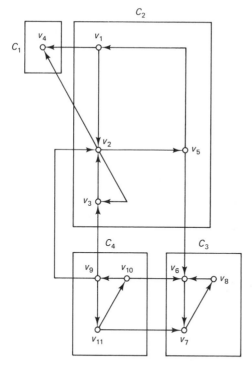

Figure 5-15. Strong components of G.

at any point CL contains fragments of many strong components.) Vertices are added to CL in tandem with SP as the search extends. Unlike SP, CL is popped only at those intermittent points at which a strong component is identified. This occurs at vertices v which satisfy the condition $Dfsnum(v) = Low(v)$ which indicates the traversal of a strong component has been completed. The vertices on CL from $Top(CL)$ up to and including v lie in the identified component and are popped.

The strong components algorithm does not explicitly use the digraph edge classification, but it is implicit. The different types of edges are treated as follows.

1. We handle tree edges in the usual manner, using them to advance the search tree. A tree edge may or may not lead to a new strong component; however, the first search edge into a strong component is always a tree edge.

2. We treat back edges as in the orientation algorithm. If (u, v) is a back edge, then u and v must be in the same strong component, since the tree path from v to u together with the back edge from u to v determine a cycle. Consequently, every vertex on this cycle is in the same strong component.

3. Forward edges have no effect on our knowledge of the strong components. If (u, v) is a forward edge, and u and v happen to lie in the same component, the value of $Low(u)$ will automatically incorporate the value of $Low(v)$, because v lies in the search subtree under u, regardless of whether (u, v) is an edge or not. On the other hand, if u and v happen to lie in different components, the algorithm will have fully explored the strong component which v lies in before it completes exploring the strong component u lies in, by virtue of the depth-first character of the traversal. Accordingly, by the time (u, v) is scanned, the vertices of v will no longer be on the CL stack. The algorithm utilizes this phenomenon by ignoring any previously reached vertices no longer on CL. Thus, as in the other case, we effectively ignore the forward edge.

4. Cross edges may lie either within a strong component (between vertices of the component) or between strong components. In the latter case, the algorithm ignores the edge in calculating *Low* (which is essentially used for intracomponent purposes) by recognizing that the terminal vertex v is not on the CL stack. On the other hand, when v is still on the CL stack, both u and v must lie in the same strong component; so the algorithm treats (u, v) like a back edge.

A formal statement of the procedure for finding strong components follows. The data types are basically the same as for the orientation algorithm. We can package an indicator array in *CL* to support efficient CL membership testing. A trace of the operation of the algorithm for the graph of Figure 5-14 is shown in Figure 5-16. The strong components are shown in Figure 5-15. Observe that only the strong components *C1*, *C2*, and *C3* are reachable from v_1.

Procedure Strong(G,v)

(* Finds and lists the strong components of G reachable from v *)

var   G: Graph
       SP, CL: Stack Entry pointer
       u, v, p: 1..|V|
       Nextcount, Top, Predecessor_of_Top: Integer function
       Next, Empty: Boolean function
       Pop: Stack Entry pointer function

Set Dfsnum(v) to Nextcount
Set Low(v)   to Dfsnum(v)
Create(SP); Push (SP,v)
Create(CL); Push (CL,v)

repeat

     (* ADVANCE *)

     while   Next(Top(SP),u)   **do**

        if   Dfsnum(u) = 0

           then   (* Advance *)
               Push(SP,u); Push(CL,u)
               Set Dfsnum(u) to Nextcount
               Set Low(u)   to Dfsnum (u)

           else   (* Compare *)
               if   u ∈ CL   **and***   Dfsnum(Top(SP)) ≥ Dfsnum(u)
               then Set Low(Top(SP)) to
                          min{Low(Top(SP)), Dfsnum(u)}

SEARCH PATH	COMPONENT LIST	ACTION	Dfsnum	Low
$v1$	$v1$	Initialization	$v1:1$	$v1:1$
$v1$	$v1$	Scan $(v1, v2)$ Push $v2$	$v2:2$	$v2:2$
$v1, v2$	$v1, v2$	Scan $(v2, v3)$ Push $v3$	$v3:3$	$v3:3$
$v1, v2, v3$	$v1, v2, v3$	Scan $(v3, v2)$		$v3:2$
$v1, v2, v3$	$v1, v2, v3$	Pop $v3$		
$v1, v2$	$v1, v2, v3$	Scan $(v2, v4)$ Push $v4$	$v4:4$	$v4:4$
$v1, v2, v4$	$v1, v2, v3, v4$	Display $\{v4\}$ and Pop $v4$ from path. Pop list up to $v4$.		
$v1, v2$	$v1, v2, v3$	Scan $(v2, v5)$ Push $v5$	$v5:5$	$v5:5$
$v1, v2, v5$	$v1, v2, v3, v5$	Scan $(v5, v1)$		$v5:1$
		Scan $(v5, v6)$ Push $v6$	$v6:6$	$v6:6$
$v1, v2, v5, v6$	$v1, v2, v3, v5, v6$	Scan $(v6, v7)$ Push $v7$	$v7:7$	$v7:7$
$v1, v2, v5, v6, v7$	$v1, v2, v3, v5, v6, v7$	Scan $(v7, v8)$ Push $v8$	$v8:8$	$v8:8$
$v1, v2, v5, v6, v7, v8$	$v1, v2, v3, v5, v6, v7, v8$	Scan $(v8, v6)$		$v8:6$
$v1, v2, v5, v6, v7, v8$	$v1, v2, v3, v5, v6, v7, v8$	Pop $v8$		$v7:6$
$v1, v2, v5, v6, v7$	$v1, v2, v3, v5, v6, v7, v8$	Pop $v7$		
$v1, v2, v5, v6,$	$v1, v2, v3, v5, v6, v7, v8$	Pop $v6$ from path. Display and Pop $\{v6, v7, v8\}$ from list.		
$v1, v2, v5$	$v1, v2, v3, v5$	Pop $v5$		$v2:1$
$v1, v2$		Pop $v2$		
$v1$	$v1, v2, v3, v5$	Pop $v1$ from path. Display and Pop $\{v1, v2, v3, v5\}$ from list.		
Empty	Empty	Exit.		

Figure 5-16. Trace of strong components algorithm from v_1.

```
(* BACK UP *)
if      Dfsnum (Top(SP)) = Low (TOP(SP))

then    (* Display strong component *)
        Pop and list CL up to and including Top(SP)

else    (* Inherit *)
        Set p = Predecessor_of_Top(SP)
        Set Low(p)   to   min {Low(p), Low(Top(SP))}
```

End_Procedure_Strong

Before establishing the correctness of the algorithm, it will be convenient to introduce the *condensation digraph D* associated with a digraph $G(V, E)$. Corresponding to each strong component s of G, we define a vertex s' in $V(D)$. There is an edge from s' to u' in $V(D)$ if and only if there is a (similarly directed) edge between the strong components corresponding to s' and u' in G. Figure 5-18 shows the condensation of the digraph of Figure 5-14.

Theorem (Correctness of Strong Components Algorithm). Let $G(V, E)$ be a directed graph, and let v be any vertex in G. Then, the strong components algorithm finds all the strong components in G reachable from v.

The proof is as follows. We will first establish the correctness of the algorithm in the special case where the digraph G is strongly connected, that is, consists of a single strong component. We will then extend the proof to the case of arbitrary digraphs. Thus, let us first assume that $G(V, E)$ is strongly connected, and let v be the root vertex used for the depth first search. We will show that in this case:

Claim: For Every Vertex $u <> v$, Low(u) < Dfsnum(u)

The proof of the claim is by contradiction. Suppose on the contrary that there exists a vertex other than v for which *Low* and *Dfsnum* are equal. Let u be the first such vertex popped from the depth-first stack, and consider the depth-first search subtree $T(u)$ rooted at u. (Refer to Figure 5-17.) Since G is strongly connected, there exists a directed path $Q(u, v)$ in G from u to v. Let w be the last vertex on $Q(u, v)$ which is contained in $T(u)$, and let z be the first vertex after w on $Q(u, v)$. Since z is not in $T(u)$, the search process must have reached z before u. Therefore, *Dfsnum*(z) must be less than *Dfsnum*(u). Furthermore, since u is the first vertex for which the *Low* and *Dfsnum* values are equal, the only condition under which the component list stack is ever popped, z must still be on the component list stack when (w, z) is scanned. Therefore, the strong

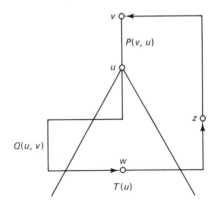

T(u): Search subtree rooted at u

P(v, u): Search path from v to u

Q(u, v): Path from u to v

w: Last vertex in $T(u) \cap Q(u, v)$

z: Successor of w on Q(u, v)

Figure 5-17. Diagram for correctness argument.

Figure 5-18. Condensation digraph for G.

connectivity procedure will set $Low(w)$ to at least as small as $Dfsnum(z)$. Since w lies in the subtree $T(u)$, the algorithm then eventually sets $Low(u)$ to at least as small as $Dfsnum(z)$, contradicting the assumption that $Low(u)$ equals $Dfsnum(u)$. The claim follows from this contradiction.

To complete the proof of the theorem in the special case, we observe that since every vertex in G is reached by the search tree before the search path retracts from v, every vertex in G must still be on the component list stack when the algorithm retracts from v and lists the members of the strong component recognized at that point. These will be precisely all the vertices in G. Thus, the algorithm correctly recognizes the sole strong component in G when G is strongly connected.

The argument in the general case of the theorem is as follows. The search starts at the vertex v in some strong component of the digraph and advances until it enters a strong component containing no outgoing edge to any other strong component. By the very same argument that we used in the special case, the algorithm will correctly recognize this strong component and remove its vertices from the component stack list. Once identified, that strong component will be subsequently ignored by the algorithm, and its vertices can never enter the component list stack again. Indeed, the search procedure acts from this point as if it were scanning a digraph lacking that strong component, and so the result follows by induction. This completes the proof of the theorem.

Refer to Figure 5-15 for an example. The search starts at component C2 and advances through strong component C3. After the algorithm has identified C3, the C3 vertices are subsequently ignored. The same procedure is repeated until every strong component reachable from the entry vertex v has been identified. Observe that the strong components are listed in the same order as a depth-first traversal of the condensation D would pop the vertices of D from a depth first stack.

5-5 BLOCK ALGORITHM

The components of a graph are its most primitive global structural feature. The blocks of a graph represent a refinement of this feature. Recall that a cut-vertex of a graph $G(V, E)$ is a vertex whose removal increases the number of components of G. A cut-vertex is also called an *articulation point* (although this term appears to be used more often when the original graph is connected). A *nonseparable* graph is a nontrivial connected graph with no articulation points. A *block* (or *bicomponent* or *biconnected component*) is a maximal nonseparable subgraph of a graph. If a block has order at least 3, it has vertex connectivity at least two. The blocks of a graph can intersect at articulation points. An articulation point always belongs to at least two blocks (possibly more), and any pair of intersecting blocks intersect at precisely one articulation point.

There are similarities and differences between components, strong components, and blocks. Thus, the components of a graph are the maximal connected parts of a graph, while the blocks of a graph are its maximal internally two vertex connected parts (except for those that happen to degenerate to a single edge). The components of a graph G partition both the vertices and edges of G into disjoint sets, while the blocks partition only the edges of G into disjoint sets, since the blocks themselves can intersect at articulation points. The strong components of a digraph, in contrast, partition the vertices, though not necessarily the edges of the digraph, since not every edge need belong to a strong component, unless it lies on a cycle.

We will present an algorithm for finding the blocks of a graph which is based on depth-first search. The algorithm is similar to the strong components algorithm. The block algorithm is based on the following theorem. The parameter *Low* referred to in the theorem is defined in the same way as the parameter *Low* used in the orientation algorithm. We leave the proof of the theorem as an exercise.

Theorem (Block/Articulation Point Structure). Let $G(V, E)$ be a graph. Let r be a vertex in G, and let $T(r)$ be the depth-first search tree rooted at r. Let a be an articulation point of G and let B be a block of G containing a. Then, all the vertices in B (except a) lie in a subtree of T rooted at a child w of a. The root r is an articulation point if and only if it has more than one child; while $a <> r$ is an articulation point if and only if it has a child w such that $Low(w) \geq Dfsnum(a)$.

The block procedure follows. The data types used in the procedure are similar to those used for the strong component and orientation algorithms. Of course, the graph G is undirected. The search path is SP. The BCL stack stores the block vertices in a nested manner and is similar to the CL stack used for the strong components algorithm. The outputs of the procedure are the sets of vertices in each block. The sets are listed as the blocks are recognized.

```
Procedure Block(G)

(* Lists the blocks of G *)

var  G: Graph
     SP, BCL: Stack Entry pointer
     u, v, p: 1..|V|
     Nextcount, Predecessor_of_Top: Integer function
     Next, Empty: Boolean function
     Top, Pop: Stack Entry pointer function

Select some v in V(G)
Set Dfsnum(v) to Nextcount
Set Low(v)    to Dfsnum(v)
Create(SP);  Push (SP,v)
Create(BCL); Push (BCL,v)

repeat

   (* ADVANCE *)

   while  Next(Top(SP),u)  do
```

```
        if      Dfsnum(u) = 0

        then    (* Tree edge − Advance *)
                Push(SP,u); Push(BCL,u)
                Set Dfsnum(u) to Nextcount
                Set Low(u)     to Dfsnum (u)

        else    (* Back edge − Compare *)
                if    Dfsnum(u) < Dfsnum(Top(SP))
                then Set Low(Top(SP)) to min{Low(Top(SP)),Dfsnum(u)}


    (* BACK UP *)

    if      Top(SP) <> v

    then    Set    p = Predecessor_of_Top(SP)

            if     Low (Top(SP)) ≥ Dfsnum (p)

            then   (* Display block *)
                   List and Pop BCL up to and including Top(SP)
                   List p

            else   (* Inherit *)
                   Set Low(p)   to   min {Low(p), Low(Top(SP))}

    until   EMPTY (POP(SP))

    End_Procedure_Block
```

For an example, refer to the diagram in Figure 5-19 which shows three blocks B1, B2, and B3 joined at an articulation point v. If the search starts at root, the block algorithm will eventually list the (vertex members of the) blocks B2 and B3, not necessarily in that order, and finally the members of block B1. The B2 block is recognized when the search retracts from w. In the search subtree $T(v)$ rooted at v, w is the only child of v in B2. All the other descendants of v in $B2$ are descendants of w. By the time B2 is scanned, $Low(w)$ has been set to $Dfsnum(v)$. The algorithm uses this phenomenon to test for the completion of the traversal of a block, and lists the appropriate vertices of

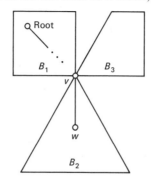

Figure 5-19. Block diagram.

B2 at this point. The same pattern recurs when we scan B3. Eventually, after both B2 and B3 have been recognized, the search returns to complete the suspended traversal of B1. B1 is recognized when the search finally retracts from the search tree child of root.

REFERENCES AND FURTHER READING

Depth-first search was introduced and applied to a variety of applications in Tarjan (1972) and (1974). Williamson (1984) uses depth-first search to extract Kuratowski subgraphs from nonplanar graphs in linear time. For a discussion of traversal techniques that avoid use of an explicit stack, such as Lindstrom's method, see Standish (1980). The alternative strong components algorithm in the exercises is from Basse (1988).

EXERCISES

1. Give an example of a class of graphs where the maximum stack length for depth-first search is $O(|V|)$, while the maximum queue length for breadth-first search is $O(1)$, and conversely.
2. Perform depth-first search and breadth-first search on $K(n)$, $C(n)$, and $K(m, n)$. What is the effect of the order of edges on the adjacency list on the traversal?
3. Compare the performance for the recursive and the nonrecursive (stack) algorithm for depth-first search on a set of test graphs; also compare the programming effort involved.
4. If a depth-first search starts at a vertex v, what is the maximum depth attained by the search stack in terms of some standard graphical invariants that depend on v? What is the minimax stack depth for the graph as a whole, over all possible starting vertices? What vertex would you start the search at in order to minimize the maximum stack depth?
5. Assume that a depth-first search tree has been constructed, and that then an edge of the graph is deleted. What happens if the search is repeated? What is the relation between the before and after search tree?
6. Show that if a vertex u in an undirected $G(V, E)$ is an ancestor (other than a parent) of v in the breadth-first search tree T on G, then (u, v) is not an edge of T. Contrast this with (undirected) depth-first search, where if (u, v) is an edge, either u or v is an ancestor of the other.
7. Adapt the breadth-first search algorithm to compute the level number of each vertex.
8. Construct a graph where breadth-first search would detect a cycle before depth-first search, and a graph where the reverse happens.
9. Contrast the performance of the breadth-first search algorithm for computing shortest paths on unweighted (that is, unit edge weight) graphs with the performance of Dijkstra's algorithm on the same graphs, as a function of the sparseness (edge density) of the graphs.
10. How do depth-first search and breadth-first search interface differently with the block structure of a graph?
11. Carefully follow the logic of the strong components algorithm as it processes a directed tree rooted at v. Check whether each step is correctly implemented.
12. Let R be the reachability matrix of a digraph $G(V, E)$. Prove $R^2(i, i)$ equals the number of vertices in the strongly connected component containing vertex i.
13. Suppose a digraph $G(V, E)$ equals its transitive closure. Design an algorithm which determines whether G is strongly connected in $O(|V|)$ time.
14. Let $G(V, E)$ be a graph. Design an algorithm that computes the minimum number of edges that have to be added to G to make it biconnected.

15. Design an algorithm based on depth-first search that computes the transitive closure of a digraph $G(V, E)$.

16. Show how to use partitioning by blocks to more efficiently find shortest distances on a graph.

17. How can partitioning using strong components be used to more efficiently find shortest distances on a digraph?

18. Adapt the orientation algorithm to the case where the input graph is mixed, that is, some of the edges are directed and some are not.

19. Prove the correctness of the block finding algorithm.

20. Prove the following algorithm finds the strong components of a digraph $G(V, E)$.

    (a) Repeat the following until every vertex of G has been labelled: perform a depth-first search of G, but number the vertices in increasing order, as they are popped from the search stack;

    (b) Reverse the directions on all the edges of G; repeat the same procedure as in step (a), restarting the search as necessary each time at the highest numbered vertex from step (a) not yet visited during this search.

    Prove that a pair of vertices are in the same strong component of G if and only if they are in the same depth-first tree constructed in step (b).

6

Connectivity And Routing

We have already defined the *vertex connectivity* VC(G) and *edge connectivity* EC(G) of a graph G. These invariants measure the susceptibility of a graph to disconnection under vertex or edge deletion. From one viewpoint, they calibrate the vulnerability of the graph considered as a communication network to disruption. It is remarkable that precisely these measures of strength of connection (or vulnerability to disconnection) also determine the diversity of routing in the graph, that is, the number of disjoint paths between pairs of vertices. After introducing some of the basic theoretical results for these invariants, we will illustrate their use in modelling and then describe certain network flow algorithms and how they can be used to calculate the vertex and edge connectivity.

6-1 CONNECTIVITY: BASIC CONCEPTS

We can define $VC(G)$ and $EC(G)$ for a graph $G(V, E)$ in terms of parametrized versions of these invariants which will be useful later in developing algorithms for VC and EC. If u and v are nonadjacent vertices in G, we define a *vertex disconnecting set (or vertex separator) between u and v* in G as a set of vertices S in $V(G)$ such that u and v are in different components of $G - S$. S is said to *disconnect* (or *separate*) the pair of vertices u and v. An *edge disconnecting set between a pair of vertices u and v* in G is similarly defined as a set of edges T in $E(G)$ such that u and v are in different components of $G - T$. For nonadjacent vertices u and v, we define the *vertex connectivity between u and v*, denoted $VC(G, u, v)$, as the cardinality of the smallest vertex disconnecting set between u and v. On the other hand, $EC(G, u, v)$ denotes the cardinality of the smallest edge disconnecting set between the pair of vertices u and v. We can then define $VC(G)$ as the minimum of $VC(G, u, v)$ taken over all distinct, nonadjacent vertices u and v in G (or $|V| - 1$ if G is a complete graph), while $EC(G)$ is the minimum of $EC(G, u, v)$ taken over all distinct vertices u and v in G. It is easy to prove:

Theorem (Bounds on Connectivity).   For every graph $G(V, E)$

$$VC(G) \leq EC(G) \leq min(G).$$

There are a variety of conditions on the degree sequence of a graph that guarantee lower bounds on connectivity, such as the following.

Theorem (Connectivity Lower Bound).   Let $G(V, E)$ be a graph of order greater than one, and let n satisfy $1 \leq n \leq |V| - 1$. If $min(G) \geq (|V(G)| + n - 2)/2$, then $VC(G) \geq n$.

By the previous theorem, this also establishes a lower bound on edge connectivity.

The precise relation between the connectivity of a graph and the number of disjoint paths that exist between pairs of vertices in the graph is the subject of the following classical theorem. Let us say a pair of paths are *vertex disjoint* if they have at most their endpoints in common; while a pair of paths are *edge disjoint*, if they have no edges in common.

Theorem (Path Diversity Characterization of Connectivity).   Let $G(V, E)$ be a graph and let k be an integer. Then,

(1) If the removal of k (but not less than k) vertices (edges) disconnects a pair of vertices u and v in G, there are at least k vertex (edge) disjoint paths in G from u to v;

(2) If G is k vertex (edge) connected, there are at least k vertex (edge) disjoint paths between every pair of vertices in G.

Figure 6-1 gives an example. Variations on this theorem guarantee vertex disjoint paths between a given vertex and a set of other vertices or between sets of vertices, as follows.

Theorem (Diverse Vertex-to-Set Routing).   If $G(V, E)$ is an n vertex connected (edge connected) graph and $v, v_1, \ldots, v_n$ are $n + 1$ disjoint vertices of G, then there are n vertex disjoint (edge disjoint) paths P_i from v to v_i, $i = 1 . . n$.

For example since the graph in Figure 6-1 has edge connectivity 2, there must be two edge disjoint paths between v_1 and $\{v_4, v_5\}$, such as the paths v_1-v_3-v_4 and v_1-v_2-v_3-v_5. More generally, we have the following.

Theorem (Diverse Set-to-Set Routing).   Let $G(V, E)$ have order at least $2n$. Then, G is n vertex connected if and only if for every pair of disjoint sets of vertices X and Y which are each of cardinality n, there exist n vertex disjoint paths between X and Y.

The proof of this theorem is as follows. First, observe that it is easy to show that if G is n vertex connected, there must be n vertex disjoint paths between X and Y. We merely add artificial vertices x and y to G which we make incident with every vertex in X and Y, respectively. The resulting graph G' is still n vertex connected. Therefore by the Path Diversity Theorem there are n vertex disjoint paths between x and y in G', which in turn determine n vertex disjoint paths between X and Y in G.

Conversely, we can show that the existence of n disjoint paths between every disjoint pair of sets of cardinality n implies G is n vertex connected. Thus, let S be a vertex disconnecting set in G which disconnects G into two parts which we denote by S_1

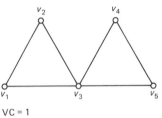

VC = 1

EC = 2

All paths between v_1 and v_5:

$$v_1 - v_2 - v_3 - v_4 - v_5$$
$$v_1 - v_3 - v_5$$
$$v_1 - v_3 - v_4 - v_5$$
$$v_1 - v_2 - v_3 - v_5$$

Pair of edge disjoint paths from v_1 to v_5:

$$v_1 - v_2 - v_3 - v_4 - v_5$$
$$v_1 - v_3 - v_5$$

Alternate pair of edge disjoint paths v_1 to v_5:

$$v_1 - v_2 - v_3 - v_5$$
$$v_1 - v_3 - v_4 - v_5$$

Maximum cardinality set of vertex disjoint paths from v_1 to v_5:

Any path from v_1 to v_5

Figure 6-1. Diverse routing example.

and S_2. $|S \cup S_1 \cup S_2|$ is at least $2n$, so we can partition S into two disjoint parts S_1' and S_2' such that $|S_1 \cup S_1'|$ and $|S_2 \cup S_2'|$ are both at least n. Define X and Y to be any subsets of $S_1 \cup S_1'$ and $S_2 \cup S_2'$, respectively, of cardinality n. By supposition, there are n vertex disjoint paths from $S_1 \cup S_1'$ to $S_2 \cup S_2'$. Since S separates S_1 from S_2, each of these paths must contain a vertex of S; hence S must contain at least n vertices. This completes the proof.

Our earlier characterization of orientable graphs implied that every pair of vertices in a (connected) bridgeless (two-edge connected) graph lie on a circuit. One can easily show for a two-vertex connected graph that every pair of vertices lie on a cycle. The following theorem generalizes this.

Theorem (Connectivity Guaranteed Cycles). If $G(V, E)$ is an n vertex connected graph, $n \geq 2$, every n vertices in G lie on some cycle.

It is nontrivial to calculate the connectivity of a graph or to identify the disjoint paths between vertices which are guaranteed by a given level of connectivity. For graphs of connectivity one, we can use depth first search to calculate both the connectivity and to identify the connecting paths. The orientation and block algorithms essentially determine if a graph has connectivity at least 2. For higher levels of connectivity, the evaluation of connectivity and the identification of disjoint paths is generally done using maximum flow algorithms, as we shall describe in a later section.

6-2 CONNECTIVITY MODELS: VULNERABILITY AND RELIABLE TRANSMISSION

Vulnerability of networks. The number of vertices or edges that must fail before a network becomes disconnected is a common measure of the *reliability* or *vulnerability* of a communication network. The corresponding graphical invariants are the vertex or edge connectivity of the network. It is natural to ask how to design a network with a given number of vertices, which attains a prescribed level of reliability, as cheaply as possible, where we measure the cost of constructing the network in terms of the number of its edges. In graph-theoretic terms, the question is how do we construct a graph $G(V, E)$ of given order and prescribed vertex connectivity, but of minimimum size?

There is an extensive literature on such extremal problems where some graphical invariants are held fixed, while others are optimized. One class of graphs that solve the minimum cost reliable network design problem just posed are the so-called *circulants,* defined as follows. Let $C(n)$ be a cycle of order n. For even values of k, the graph $C(n)^{(k/2)}$, has connectivity k, and the minimum number of edges $(nk/2)$ among all graphs of connectivity k and order n.

Reliable transmission (Byzantine generals problem). Consider a distributed system of processors whose interconnection network is given by an undirected graph $G(V, E)$. Assume there is a processor at each vertex of G which can communicate directly with the neighboring processors at adjacent vertices. To transmit a message to a distant processor, a processor includes the intended route with its message and then routes the message through the network, by transmitting it to the first processor/vertex on the route, which proceeds in a similar manner, and so on, until the message eventually reaches its destination.

We are interested in the kind of problems that can arise when some of the processors in the network may be faulty. We will assume that faulty processors can exhibit a bizarre variety of deleterious effects. They may lose messages, route messages incorrectly, garble messages, or even generate multiple incorrect copies of messages. It is not at all obvious whether reliable transmission can be achieved in the presence of such potential chaos. But, surprisingly, we can show that if the interconnection graph for the network has sufficiently high vertex connectivity, then the reliable processors in the network can communicate reliably even in the presence of severely unreliable components (Dolev (1982)).

It will be convenient to introduce the following terminology. Let S be a set of vertex disjoint paths in $G(V, E)$ from a reliable processor T to a reliable processor P. We will define a path in S to be *good* if every processor on the path is reliable. We define a path in S to be *bad* if there is at least one faulty processor on the path. Let F be an upper bound on the number of faulty processors in G. A *suspicious set of paths from T to P* is a set U of at most F vertex disjoint paths in G with the property that all messages from T to P which are not routed through U arrive at P with the same message value. Refer to Figure 6-2 for an illustration.

Before describing the reliable routing procedure, let us specify some assumptions.

(1) Each reliable processor can always correctly identify that adjacent processor from which it receives a given message even when the adjacent processor is unreliable.

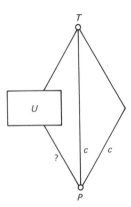

Figure 6-2. Messages not using U are identical.

(2) Each reliable processor knows the global definition of the communication network graph G and the sets of disjoint paths used by each reliable processor for processor to processor communication.

Under these assumptions we can prove the following theorem.

Theorem (Reliable Transmission Condition). Let $G(V, E)$ be a graph model of a network with at most F faulty processors. Let the vertex connectivity of G be at least $2F + 1$. Suppose T is a reliable transmitter in G which sends $2F + 1$ copies of a message m to a reliable processor P through $2F + 1$ known vertex disjoint paths. Then, P can correctly receive m by finding a set of suspicious paths U from T to P and selecting the unique message value not routed through U.

The proof of the theorem is as follows. Let M denote the set of messages received at P. Some of these are correct copies of the original message m and some are invalid copies. P must try to determine which represent(s) the correct message. Each message contains a description of the path it took from T to P. Of course, the description may have been perverted by a faulty processor along that route, and the message itself could be spurious. Nonetheless, since the $2F + 1$ paths T uses to transmit to P are known (by assumption (2)) both to P and to all the other reliable processors in the network, the reliable processors can ensure in the first place that incorrect messages are restricted to bad paths. For example, if an unreliable processor attempts to move an incorrect message from a bad path to a good path, the first (reliable) processor on the good path to receive the message will recognize (by assumption (1)) that the route is an invalid T to P route and so can delete the message. It follows similarly that the routing description delivered with an incorrect message must be for a bad path, though not necessarily the bad path that was actually taken.

There are at most F bad paths from T to P since there are no more than F faulty processors. Since T uses $2F + 1$ disjoint paths from T to P, at least $F + 1$ valid copies of the message transmitted must reach P. An even larger number of invalid messages can also arrive at P, but P can filter out the invalid messages by the following technique of identifying a suspicious set of paths U from T to P.

Let us first observe that some suspicious sets of paths certainly exist. For example, if we set U to the set of disjoint paths that actually contain faulty processors, U is a suspicious set since there are at most F such paths and any messages not passing through U will have the same (correct) value.

We can show that the common message value transmitted not using a suspicious set U is always the correct message value. To prove this, observe that U contains at most F good paths. Since there are at least $F + 1$ good paths, there is at least one good path not in U. Therefore, at least one correct copy of the message is routed outside U. By the definition of a suspicious set, only a single message value is routed outside U. Therefore, that value must always be the correct one. This completes the proof of the theorem.

6-3 NETWORK FLOWS: BASIC CONCEPTS

A *network* is a digraph $G(V, E)$ with nonnegative, real-valued weights assigned to each edge. Alternatively, a network is an ordered triple (V, E, w), where V is a set of vertices, E a set of directed edges on V, and w is a function mapping the elements of E to the nonnegative reals. If we interpret the edge weights as limits on the capacities of the edges to transmit a commodity of some type (such as, vehicular traffic, data, fluids), the network is called a *capacitated network*. We will show how to find the maximum amount of traffic that can be transmitted between a given pair of vertices in a capacitated network. This is called the maximum network flow problem. We present two algorithms for finding the maximum flow: the classical one of Ford and Fulkerson, and a more efficient but more complicated algorithm based on the work of Dinic. The Ford-Fulkerson algorithm uses a search tree to iteratively build up the flow in the network; while the Dinic algorithm uses a "layered network" for the same purpose.

Before proceeding with a description of the maximum flow algorithms, we will need to introduce some terminology, beginning with a precise definition of what we mean by a traffic flow. There are two ways to define this, either by a path model of flow, or using conservation equations. While the two methods are mathematically equivalent, the path model is more intuitive, while the conservation model is mathematically more convenient. We give each definition, and then prove their equivalence.

Path model of network flow. Let $N(V, E)$ be a capacitated network. Let S and T be a pair of distinct vertices in $V(N)$, and let P be a set of paths from S to T. If x is an edge in $E(N)$, we denote the (nonnegative integral) capacity of x by $Cap(x)$, and we denote the set of paths in P that include the edge x by $P(x)$. Let f be a nonnegative integral function defined on P. We denote the sum of the values of $f(p)$ taken over all paths p in $P(x)$ by $Sum(f, x)$. We call f a *flow on* P if for every edge x in $E(N)$, $Sum(f, x)$ is at most $Cap(x)$. If f is a flow on P and p is a path in P, we call $f(p)$ the *flow on* p. If x is an edge in $E(N)$, we call $Sum(f, x)$ the *flow on* x. The set of paths P together with a flow f on P is called a *path flow on* N and is specified by the quintuple (N, S, T, P, f). The *value of a path flow* (N, S, T, P, f) is the sum of the values of $f(p)$ taken over all paths p in P and is denoted by $Value(P, f)$, the other parameters being left implicit. The vertices S and T are called the *source* and *sink* of the path flow, respectively.

Figure 6-3 gives an example. S and T are v_1 and v_4 respectively, the paths P are given by $\{v_1\text{-}v_2\text{-}v_4, v_1\text{-}v_3\text{-}v_4, v_1\text{-}v_2\text{-}v_3\text{-}v_4\}$, and $f(p)$ is 1 for each path. $Sum(f, x)$ is given for each edge in Figure 6-3b and is within capacity constraints. $Value(P, f)$ equals 3. We obtain a different path flow if we set P equal to $\{v_1\text{-}v_2\text{-}v_3\text{-}v_4\}$ and $f(p)$ equal to 2. $Value(P, f)$ equals 2, which we know from the previous example is less than the maxi-

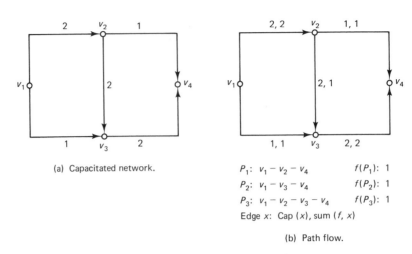

(a) Capacitated network.

P_1: $v_1 - v_2 - v_4$ $f(P_1)$: 1
P_2: $v_1 - v_3 - v_4$ $f(P_2)$: 1
P_3: $v_1 - v_2 - v_3 - v_4$ $f(P_3)$: 1
Edge x: Cap (x), sum (f, x)

(b) Path flow.

Figure 6-3. Path flow example.

mum path flow value on N. Interestingly, we cannot augment the value of this flow by adding additional flow paths to P because the existing path prevents this. A path flow with this characteristic is called unaugmentable or *maximal*. The existence of maximal but not maximum value path flows demonstrates how the resources of a network can be squandered by routing traffic carelessly.

Conservation model of network flow. The *conservation model* interprets the traffic flow on a network as composed of edge flows which are subject to conservation constraints at the vertices of the network, rather than as a superposition of flow paths. We will see this is mathematically and algorithmically more convenient to work with than the path flow model.

We introduce the conservation model with an example. Consider the path flow in Figure 6-3b. If we denote the sum of the flows on the flow paths entering a vertex v by *Path-inflow(v)*, and the sum of the flows on the flow paths exiting a vertex v by *Path-outflow(v)*, the following relations hold:

(1) *Path-inflow(S)* $= 0$, *Path-outflow(S)* $= 3$,
(2) *Path-inflow(T)* $= 3$, *Path-outflow(T)* $= 0$,
(3) *Path-inflow(v_1)* $=$ *Path-outflow(v_1)* $= 2$, and
    Path-inflow(v_2) $=$ *Path-outflow(v_2)* $= 2$.

Thus, *Path-Inflow* and *Path-Outflow* are balanced at every vertex other than the source S and the sink T. Conversely, the *net flow out of S* (defined as *Path-outflow(S)* $-$ *Path-inflow(S)*) equals the *net flow into T* (defined as *Path-inflow (T)* $-$ *Path-outflow(T)*) and each equals 3, the value of the path flow.

These relations are typical and suggest the following definitions. Let $N(V, E)$ be a capacitated network and let S and T be a pair of vertices in N. Let f be a nonnegative integral function on $E(N)$. We denote the sum of the values of f on incoming edges at a vertex v in $V(N)$ by *Inflow(v)*. *Outflow(v)* is defined similarly. We call f an *edge flow on N* if

(1) For every edge x in $E(N)$, $f(e) \le Cap(x)$.

(2) For every vertex $v <> S$ or T, $Inflow(v) = Outflow(v)$.

Condition (1) is called a *capacity constraint,* while (2) is called a *conservation constraint.* We define a *conservative flow on N from S to T* as a quadruple (N, S, T, f) where f is an edge flow on N. If x is an edge in $E(N)$; then $f(x)$ is called the *edge flow on x.* The *value of a conservative flow* (N, S, T, f) is the net flow into T, $Inflow(T) - Outflow(T)$, and is denoted by $Conval(f)$. We denote the value of a maximum possible value conservative flow from S to T on a capacitated network N by $maxflow(N, S, T)$.

We will prove that any conservative flow can be interpreted as a superposition of path flows of equivalent flow value, and conversely. But, first we need to establish some basic bounds on the value of a flow.

Maximum network flows and minimum capacity cuts. The fundamental bound on the maximum value of a path flow or a conservative flow on a network N with source S and sink T can be informally stated as follows: The maximum flow value is bounded above by the tightest "bottleneck" between S and T. We now formalize this intuitive notion.

Let (N, S, T, f) be a conservative flow. If X is a set of vertices in $V(N)$, then X^c denotes the complement of X with respect to $V(N)$, that is, $V(N) - X$. We define a *cut* (X, X^c) as the set of edges $\{(x, x') \mid x \in X$ and $x' \in X^c\}$. If S is in X and T is in X^c, then (X, X^c) is called an *S-T cut.* If (X, X^c) is an *S-T* cut, there is no path from S to T in $N(V, E-(X, X^c))$. We define the *capacity of a cut,* denoted by $Cap(X, X^c)$, as the sum of the capacities of the edges in (X, X^c). We denote the capacity of a minimum capacity *S-T* cut in N by $mincap(N, S, T)$. The *flow on* (X, X^c), denoted $flow(X, X^c)$, is the sum of the edge flows on edges in (X, X^c). The *net flow across* (X, X^c), denoted by $Netflow(X, X^c)$, is $flow(X, X^c) - flow(X^c, X)$. Refer to Figure 6-4 for an illustration.

Theorem (Net Flow and Flow Value). Let (N, S, T, f) be a conservative flow and let (X, X^c) be an *S-T* cut in N. Then,

$$flow(X, X^c) - flow(X^c, X) = Conval(f).$$

The proof of the theorem is as follows. For every vertex v in X^c other than T, the following conservation equations hold:

$$\sum_x f(x, v) - \sum_y f(v, y) = 0$$

where the first sum is taken over all vertices x in $V(N)$ adjacent to v, while the second sum is taken over all vertices y in $V(N)$ adjacent from v. At the sink T, we have

$$\sum_x f(x, T) - \sum_y f(T, y) = Conval(f),$$

where x and y are defined similarly. The sum of the right hand sides taken over all these equations is $Conval(f)$, while we can show the sum of the left hand sides taken over all the equations equals $flow(X, X^c) - flow(X^c, X)$. For, any edge (x, y) with x and y both in X^c, contributes opposite terms to the sum of the left hand sides. The edge

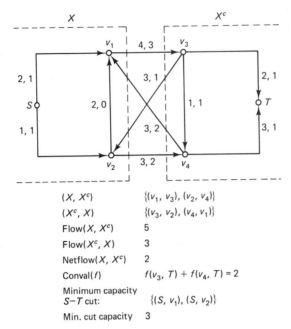

(X, X^c)	$\{(v_1, v_3), (v_2, v_4)\}$
(X^c, X)	$\{(v_3, v_2), (v_4, v_1)\}$
Flow(X, X^c)	5
Flow(X^c, X)	3
Netflow(X, X^c)	2
Conval(f)	$f(v_3, T) + f(v_4, T) = 2$
Minimum capacity S–T cut:	$\{(S, v_1), (S, v_2)\}$
Min. cut capacity	3

Figure 6-4. (N, S, T, f) cut example.

(x, y) generates a positive term $+f(x, y)$ considered as an incoming edge at y and an opposite negative term $-f(x, y)$ when considered as an outgoing edge at x; so that the terms of this form cancel. The only remaining terms arise from flows on edges in (X, X^c) (positive terms) and flows on edges in (X^c, X) (negative terms); so the equations overall sum to

$$flow(X, X^c) - flow(X^c, X) = Conval(f),$$

which completes the proof of the theorem.

The next theorem follows readily from the Net-Flow Theorem. It shows that the value of the maximum possible S to T flow realizable on a given capacitated network is never greater than the capacity of the smallest cut S-T on the network. We will use this theorem later in establishing both the termination and correctness of the maximum flow algorithms. We will also see that the upper bound in the theorem is always attainable, that is, the maximum flow value equals the minimum cut capacity.

Theorem (Maximum Flow/Minimum Capacity Cut Bound). Let $N(V, E)$ be a capacitated network and let S and T be distinct vertices in N. Then,

$$maxflow(N, S, T) \leq mincap(N, S, T).$$

The proof of this theorem is as follows. Observe that for any conservative flow (N, S, T, f) and any S-T cut (X, X^c)

$$flow(X, X^c) \leq Cap(X, X^c),$$

by the capacity constraints on a flow. Consequently,

$$Conval(f) = Netflow(X, X^c)$$
$$= flow(X, X^c) - flow(X^c, X)$$
$$\leq Cap(X, X^c),$$

by the Net Flow theorem and the nonnegativity of $flow(X^c, X)$. Consequently, the value of any conservative flow (N, S, T, f) is bounded above by the capacity of every S-T cut. Hence, the maximum flow value is bounded by the minimum cut capacity. This completes the proof of the theorem.

Equivalence of path and conservation models. We will now show that the path and the conservative flow models are equivalent in the sense that given any path flow there is a conservative flow of equal value, and conversely.

Theorem (Conservative and Path Flow Equivalence). If (N, S, T, P, f') is a path flow, there exists a conservative flow (N, S, T, f) such that $Value(f')$ equals $Conval(f)$. Conversely, if (N, S, T, f) is a conservative flow, there exists a path flow (N, S, T, P, f') such that $Conval(f)$ equals $Value(f')$.

The proof is as follows. To convert a path flow (N, S, T, P, f') into a conservative flow (N, S, T, f) of the same value, we define for each edge x in $E(N)$ a function $f(x)$ equal to $Sum(f', x)$. The function f is nonnegative and satisfies both the capacity and conservation of flow constraints at vertices other than S or T which is the definition of an edge flow. The quadruple (N, T, S, f) then defines a conservative flow such that $Conval(f)$ equals $Inflow(T)$ which equals $Value(f)$ as required.

We can convert a conservative flow (N, S, T, f) into a path flow (N, S, T, P, f') of the same value by repeatedly identifying and "removing" S to T paths in the conservative flow network all of whose edges have positive flow. Whenever we find a path, we assign it a path flow value equal to the minimum of the edge flows on the path, and we reduce the edge flows along the path by the value of the path flow. We repeat the procedure until we can no longer find a nonzero flow path at which point the total value of the path flows removed equals the value of the original conservative flow.

To prove that we can repeat this process until the value of the conservative flow has been reduced to zero, we argue as follows. If the conservative flow has nonzero flow value, we can prove by the Net-Flow Theorem that there exists a flow path of nonzero value. We let X denote the set of vertices in $V(N)$ reachable from S by paths containing only edges with nonzero edge flows. If T is in X, we can certainly find a nonzero path flow from S to T. On the other hand, if T is not in X, then (X, X^c) is an S-T cut and all the edges in the cut have zero edge flows. Therefore, $flow(X, X^c)$ is zero whence by the Net-Flow Theorem and the nonnegativity of the edge capacities, $flow(X^c, X)$ is also zero. Thus, the conservative flow already has value zero, which completes the proof of the theorem.

A depth first search procedure for extracting flow paths Get_Flow_Path is given below. The procedure tries to find a path from S to T containing only edges with nonzero edge flows which thus defines a nonzero flow path. We can call Get_Flow_Path repeatedly until $Outflow(S)$ is reduced to zero. The final reduced flow network has zero flow value, although the flow mapping may not be identically zero because "circulatory" flows not contributing to any S-T flow may remain.

We represent the capacitated network $N(V, E)$ as a linear array with vertices and edges of type

```
type  Vertex  =  record
                 Cap, Path-Val: Nonnegative Integer
                 Dfs, Pred: 0..|V(N)|
```

```
                        Positional-Pointer, Successor: Edge pointer
                    end


        Edge  =     record
                        Nghb: 1..|V(N)|
                        f (flow): Nonnegative Integer
                        Edge-successor: Edge pointer
                    end
```

The meaning of most of the fields is clear from the field names. Path-Val(v) stores the value of the flow on the depth first flow path to v. Pred(v) points to the predecessor of v in the depth first tree. Dfs is the field that would usually contain the depth first number of the vertex; however, we use it as a flag which indicates whether the vertex has been visited yet during the current invocation of Get_Flow_Path. The source and sink are distinct vertices S and T in $V(N)$. The depth first stack is denoted by ST. The flow path (or at least its successive vertices in reverse order) is returned in ST, while the value of the flow path is returned in Val. M denotes some large positive integer which plays the role of plus infinity.

We denote the conditional **and (or)** operator by **and* (or*)**. The second condition in A **and*** B (A **or*** B) is not tested if the first condition fails (succeeds). The auto-increment function Nextcount is initially zero, as are the Dfs fields. A global counter k is incremented by one at each invocation of Get_Flow_Path. When we consider the maximum flow algorithm in the next section, we will need to enhance the network representation so as to facilitate access to both adjacent to and adjacent from vertices, since the maximum flow algorithm considers both incoming and outgoing edges at each vertex. But, for our present purposes, we only need adjacent to vertices, so the function Next(x, y) may be defined as usual.

We refer to Figure 6-5 for an example of the procedure. The final conservative flow in the example has value zero, though it is not identically zero.

```
        Function Get_Flow_Path (N, S, T, ST, Val)


        (* This returns an S to T flow path in ST with value Val,
            reducing the conservative flow (N,S,T,f) accordingly, and
            fails if there is none. *)


        var  N: Network
             S,T,u,x,p: 1..|V|
             k, Val: Nonnegative Integer
             M: Integer Constant
             Next, Empty, Get_Flow_Path: Boolean function
             Nextcount,Top: Integer function
             ST: Stack pointer
             Pop: Stack pointer function


        Set k to Nextcount
        Set Dfs(S) to k
        Set Path-Val(S) to M
        Create(ST); Push (ST,S)
```

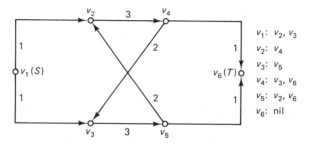

(a) Conservative flow with edge flows shown.

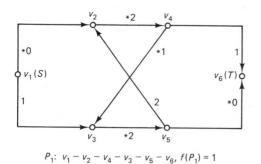

P_1: $v_1 - v_2 - v_4 - v_3 - v_5 - v_6$, $f(P_1) = 1$

(b) Conservative flow after removal of flow path P_1.

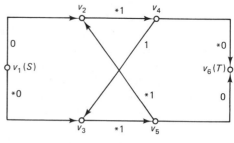

P_2: $v_1 - v_3 - v_5 - v_2 - v_4 - v_6$, $f(P_2) = 1$

(c) Conservative flow after removal of second flow path P_2.

Figure 6-5. Application of Get_Flow_Path.

(* Try to find an S to T flow path *)

repeat

    while TOP(ST) <> T **and*** Next (TOP(ST), u) **do**

        if   Dfs (u) < k **and** f(TOP(ST),u) > 0
        then **Set** Dfs (u) to k
               Set Pred(u) to Top(ST)
               Push(ST,u)
               Set Path-Val(u) to min{Path-Val(Top(ST)),f(TOP(ST),u)}

    until TOP(ST) = T **or*** EMPTY (POP(ST))

    (* If non-zero flow path found, reduce affected edge flows *)

    Set Get_Flow_Path to **not** EMPTY(ST)

    if   Get_Flow_Path

    then (* Reduce flow along path *)
            Set Val to Path-Val(T)
            Set x to Top(ST)

```
        while x <> S do
            Set p to Pred(x)
            Set f(p,TOP(ST))  to f(p,TOP(ST)) − Path-Val(T)
            Set x to p

    End_Function_Get_Flow_Path
```

6-4 MAXIMUM FLOW ALGORITHM: FORD AND FULKERSON

The maximum flow algorithm of Ford and Fulkerson uses generalized flow paths to modify the values of the edge flows of a conservative flow in a way that both preserves the conservative nature of the flow and increases the value of the flow. The "paths" are called flow augmenting paths and not only provide a means of increasing the value of a flow, but also allow us to characterize maximum flows algorithmically.

We will require some preliminary definitions. Let (N, S, T, f) be a conservative flow. We say an edge x in $E(N)$ is *saturated* with respect to (N, S, T, f) if $f(x)$ equals $cap(x)$. We define the *spare capacity* of an edge x as the difference $cap(x) - flow(x)$. The simplest kind of flow augmenting path is a directed path from S to T that contains no saturated edges. If we denote the minimum of the spare capacities taken over all the edges of such a path by *Inc* and increment the edge flow on every edge of the path by *Inc*, the resulting flow has flow value $Conval(f) + Inc$.

It may happen that the flow value $Conval(f)$ is less than $maxflow(N)$ even when there are no such directed paths along which we can increase the flow. Figure 6-6 gives an example. However, we can define flow augmenting paths of such generality that their absence guarantees the flow value is maximum.

We define a *flow augmenting path* (FAP) with respect to a conservative flow (N, S, T, f) as an alternating sequence of vertices and edges of $N(V, E)$:

$$v_1, e_1, \ldots v_i, e_i, \ldots, v_n$$

such that

(1) v_1 equals S and v_n equals T.
(2) Each edge e_i $(i = 1, \ldots, n - 1)$ is either of the form (v_i, v_{i+1}) or of the form (v_{i+1}, v_i) and is called respectively a *forward* or a *backward* edge of the sequence.
(3) Every forward edge e_i $(i = 1, \ldots, n - 1)$ of the sequence has positive spare capacity, that is

$$cap(e_i) - flow(e_i) > 0.$$

(4) Every backward edge e_i $(i = 1, \ldots, n - 1)$ has positive flow, that is

$$flow(e_i) > 0.$$

We refer to Figure 6-7 for an example. The Ford and Fulkerson maximum flow algorithm is essentially a search tree technique for finding flow augmenting paths.

In order to increase the flow value using flow augmenting paths, we let M denote the minimum of the spare capacities on the forward edges of the FAP and let m denote the minimum of the edge flows on the backward edges of the FAP. Denote $min(M, m)$ by *Inc*. We then change the edge flows along the FAP by increasing the edge flow on each forward edge by *Inc* and by decreasing the edge flow on each backward edge of

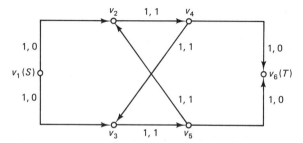

Figure 6-6. Network with no simple flow augmenting path.

the FAP by *Inc*. The resulting conservative flow has flow value *Conval*(f) + *Inc*. Refer to Figure 6-7 for an example where Inc equals *min*(3, 2) and also to Figure 6-8.

We can characterize the maximum value flow in terms of flow augmenting paths as follows.

Theorem (Flow Augmenting Path Characterization of Maximum Flow). Let (N, S, T, f) be a conservative flow. Then, *Conval*(f) equals *maxflow*(f) if and only if there is no flow augmenting path with respect to (N, S, T, f).

The proof of the theorem will follow from the proof of correctness of the maximum flow algorithm.

Maximum flow algorithm. The Ford-Fulkerson algorithm uses procedure Flow_Augmenting_Path to find FAPs. The procedure constructs a search tree consisting of partial flow augmenting paths. If the search tree reaches T, the algorithm has

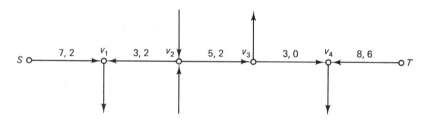

Figure 6-7. A flow augmenting path.

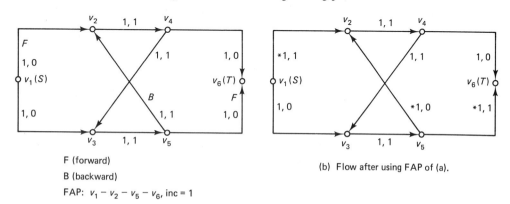

F (forward)

B (backward)

FAP: $v_1 - v_2 - v_5 - v_6$, inc = 1

(a) General FAP for network of Figure 6-6.

(b) Flow after using FAP of (a).

Figure 6-8. Flow augmentation using FAP.

found a flow augmenting path and the edge flows along the path are then changed, resulting in a greater flow value. On the other hand, if the search tree is blocked before reaching T, the blocked tree determines a cut of minimum capacity, and we can conclude that the flow value is a maximum. The maximum flow algorithm merely consists in calling Flow_Augmenting_Path repeatedly until it fails to find a further flow augmenting path.

We use breadth first search in Flow_Augmenting_Path because it can be shown (Edmonds and Karp (1972)) that if the search algorithm uses breadth first search and a shortest augmenting path, then in that case we never need more than $O(|V||E|)$ FAPs to obtain the maximum flow. Flow_Augmenting_Path is easily seen to be $O(|E|)$, so the overall performance of this algorithm is then $O(|V||E|^2)$. Figure 6-9 illustrates a problem that can arise if other search techniques are used. Thus, if the search alternates between the FAPs, S-v_2-v_3-T and S-v_3-v_2-T, it takes $2M$ calls to Flow_Augmenting_Path before the maximum value flow is found; thus making the performance of the algorithm dependent on the (minimum) capacity of the network. This phenomenon is avoided by breadth first search.

We will represent the capacitated network $N(V, E)$ as a linear array. Since we need to access both forward and backward edges, we use a shared representation for the edges. Thus, if directed edges (u, v) and (v, u) are incident with the vertices u and v, then the shared representation, which is pointed to by the edge entries on the adjacency sublists for both u and v, will contain the capacity and flow information for both edges. The vertex and edge types are as follows.

```
type Vertex =   record
                    Path-Val: Nonnegative Integer
                    Bfs, Pred: 0..|V(N)|
                    Positional-Pointer,
                        Successor, Temp: Edge pointer
                end


     Edge   =  record
                    Shared-rep: Shared-edge pointer
                    Edge-successor: Edge pointer
                end


     Shared-edge = record
                        E(2,2): 0..|V|
                        cap(2), flow(2): Nonnegative Integer
                    end
```

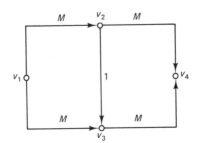

Figure 6-9. Dependency of performance on search technique.

Connectivity and Routing   Chap. 6

Path-Val(v) stores the increment attainable on the partial flow augmenting path from S to v. We use Bfs to indicate whether a vertex has been scanned yet during the current invocation of Flow_Augmenting_Path. Bfs should be cleared prior to invoking Flow_Augmenting_Path. Pred(v) points to the search path predecessor of v. If the successor of v on the FAP is u, Temp(v) points to the entry for u on the edge list of v. This facilitates updating the edge flows on the FAP when one is found. Shared-rep points to the descriptor record for the edge. The vector $(\mathbf{E(i, 1)}, \mathbf{E(i, 2)})$, $i = 1 . . 2$, gives the index of the initial and terminal vertex of edge i and is zero if the edge does not exist. Thus, if (u, v) and (v, u) are both edges, $(\mathbf{E(1, 1)}, \mathbf{E(1, 2)})$ equals (u, v) and $(\mathbf{E(2, 1)}, \mathbf{E(2, 2)})$ equals (v, u). The fields cap(i) and flow(i) give the capacity and flow of the ith edge.

We denote the breadth first queue by Q. We use a function Next(x,y,y-edge-ptr) to return in y the next neighbor of x together with a pointer in y-edge-ptr to the edge entry for y on the adjacency sublist of x, or fail. We consider the vertex y a neighbor of x if either (x, y) or (y, x) (corresponding to a forward or backward edge, respectively) is an edge. We also use a function Try_to_Augment(x,y,y-edge-ptr) which determines whether the edge (x, y) or (y, x) is augmentable. The function Nextcount is autoincrementing, starts at zero, and should be cleared prior to invoking Flow_Augmenting_Path.

Function Flow_Augmenting_Path (N,S,T)

(* This tries to find and apply a FAP from S to T, and fails if there is none *)

var N: Network
     S,T,u,x,y: 1..|V|
     u-edge-ptr: Edge pointer
     k: Nonnegative Integer
     M: Integer Constant
     Next, Empty: Boolean function
     Flow_Augmenting_Path, Try_to_Augment: Boolean function
     Head: Integer function
     Q: Queue pointer
     Dequeue: Queue pointer function

Set Path-Val(S) to M
Enqueue (Q, S)

(* Try to find a FAP from S to T *)

repeat

    while Head(Q) <> T **and*** Next (Head(Q), u, u-edge-ptr) **do**

        if   Bfs(u) = 0 **and***
                    Try_to_Augment (Head(Q),u,u-edge-ptr)

        then Set Bfs(u) to 1
              Enqueue(Q,u)

```
until   Head(Q) = T  or*  Empty (Dequeue(Q))

(* Augment edge flow along FAP, if one was found *)

Set   Flow_Augmenting_Path to not Empty (Q)

if    Flow_Augmenting_Path

then  Set x to Head(Q)
      repeat
          Set   y to Pred(x)
          if    cap(x,y) − f(x,y) > f(y,x) (using Temp(y))
          then  Set f(x,y) to f(x,y) + Path-Val(T)
          else  Set f(y,x) to f(y,x) − Path-Val(T)
          Set x to y
      until x = S


End_Function_Flow_Augmenting_Path

Function Try_to_Augment (x, y, y-edge-ptr)

(* This tries to augment the flow on the forward or backward edges
   at x which are referenced via y-edge-ptr *)

var   N: Network
      Inc: Nonnegative Integer
      x: 1..|V|
      y-edge-ptr: Edge pointer
      Q: Queue pointer

Set Inc to max (cap(x,y) − f(x,y), f(y,x))
Set Try_to_Augment to (Inc > 0)

if Inc > 0  then  Set Pred(y) to x
                  Set Temp(x) to y-edge-ptr
                  Set Path-Val(y) to min (Path-Val(x), Inc)


End_Function_Try_to_Augment
```

We refer to Figures 6-10 and 6-11 for an example. Observe how the forward edges of the cut determined by the final blocked search tree in Figure 6-10d are saturated. The vertices in the blocked search tree define a cut $\{(v_2, v_4), (v_3, v_4)\}$ of capacity 4 equal to the maximum flow value.

Theorem (Correctness of Ford And Fulkerson Algorithm).   Let $N(V, E)$ be a capacitated network with distinct vertices S and T and let (N, S, T, f) be the conservative flow determined by the Ford and Fulkerson algorithm. Then, the flow value $Conval(f)$ has a maximum possible value among all conservative flows on N.

The proof of this theorem is as follows. We first show that the algorithm terminates, and then that on termination the flow must have maximum value. To establish termination, observe that each invocation of Flow_Augmenting_Path increases the

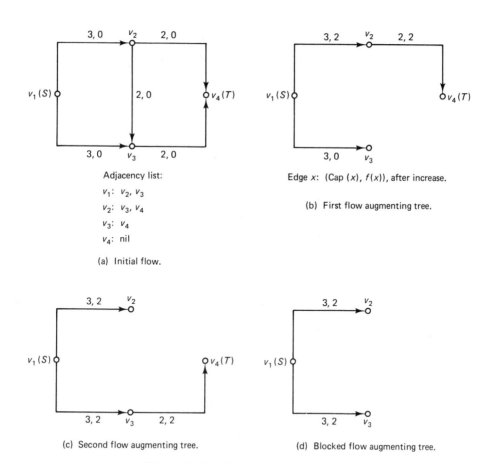

(a) Initial flow.

Adjacency list:

v_1: v_2, v_3

v_2: v_3, v_4

v_3: v_4

v_4: nil

Edge x: (Cap (x), f(x)), after increase.

(b) First flow augmenting tree.

(c) Second flow augmenting tree.

(d) Blocked flow augmenting tree.

Figure 6-10. Ford-Fulkerson example.

flow value by an integral amount Path-Val(T). Since $maxflow(N)$ is bounded above by $mincap(N)$, the algorithm must terminate within $mincap(N)$ calls to Flow_Augmenting_Path.

We show next that the value of the final flow equals $mincap$(N) and so must equal $maxflow(N)$ by the Maximum Flow / Minimum Capacity Cut Theorem. Let us denote the blocked search tree by BT. Then, $(V(BT), V(N) - V(BT))$ is an S-T cut. Every edge of the cut is saturated since otherwise some edge would be returned by Try_to_Augment. Similarly, every edge of $(V(N) - V(BT), V(BT))$ must have zero flow since otherwise some edge would be returned by Try_to_Augment. It follows from the Net-Flow Theorem that

$$\text{Conval}(f) = \text{flow}(X, X^c) - \text{flow}(X^c, X)$$

$$= \text{Cap}(X, X^c) - 0.$$

Thus, the flow value equals the capacity of an S-T cut. Since $maxflow(N, S, T) \le mincap(N, S, T)$, (N, S, T, f) must be a maximum flow and the cut must be a minimum capacity S-T cut. This completes the proof of the theorem.

The following theorem follows directly from the previous proof and summarizes the relation between maximum flows and minimum cuts.

Flow _Augmenting _Path (First invocation)

QUEUE	TREE	ACTION	AMOUNT	REMARKS
v_1	v_1	Initialize	$v_1:M$	
v_1	v_1	Scan (v_1, v_2)	$v_2:3$	
		Enqueue(v_2)		
		Add v_2 to Tree		
v_1, v_2	v_1, v_2	Scan (v_1, v_3)	$v_3:3$	
		Enqueue(v_3)		
		Add v_3 to Tree		
v_1, v_2, v_3	v_1, v_2, v_3	Dequeue(Q)		
v_2, v_3	v_1, v_2, v_3	Scan (v_2, v_3)		v_3 in Tree
v_2, v_3	v_1, v_2, v_3	Scan (v_2, v_4)	$v_4:2$	Breakthrough
		Enqueue(v_4)		
		Add v_4 to Tree		

Increase flow on v_1-v_2-v_4 by 2.

Flow _Augmenting _Path (Second invocation)

QUEUE	TREE	ACTION	AMOUNT	REMARKS
v_1	v_1	Initialize	$v_1:M$	
v_1	v_1	Scan (v_1, v_2)	$v_2:1$	
		Enqueue (v_2)		
		Add v_2 to Tree		
v_1, v_2	v_1, v_2	Scan (v_1, v_3)	$v_3:3$	
		Enqueue (v_3)		
		Add v_3 to Tree		
v_1, v_2, v_3	v_1, v_2, v_3	Dequeue(Q)		
v_2, v_3	v_1, v_2, v_3	Scan (v_2, v_3)		v_2 in Tree
v_2, v_3	v_1, v_2, v_3	Scan (v_2, v_4)		Edge saturated
v_2, v_3	v_1, v_2, v_3	Dequeue(Q)		
v_3	v_1, v_2, v_3	Scan (v_3, v_4)	$v_4:2$	Breakthrough
		Enqueue (v_4)		
		Add v_4 to Tree		

Increase flow on v_1-v_3-v_4 by 2.

Flow _Augmenting _Path (Third invocation)

QUEUE	TREE	ACTION	AMOUNT	REMARKS
v_1	v_1	Initialize	$v_1:M$	
v_1	v_1	Scan (v_1, v_2)	$v_2:1$	
		Enqueue (v_2)		
		Add v_2 to Tree		
v_1, v_2	v_1, v_2	Scan (v_1, v_3)	$v_3:1$	
		Enqueue (v_3)		
		Add v_3 to Tree		

v_1, v_2, v_3	v_1, v_2, v_3	Dequeue(Q)	
v_2, v_3	v_1, v_2, v_3	Scan (v_2, v_3)	v_3 in Tree
v_2, v_3	v_1, v_2, v_3	Scan (v_2, v_4)	Edge saturated.
v_2, v_3	v_1, v_2, v_3	Dequeue(Q)	
v_3	v_1, v_2, v_3	Scan (v_3, v_4)	Edge saturated.
v_3	v_1, v_2, v_3	Dequeue(Q)	
Empty			Search blocked

Figure 6-11. Trace of successive calls to Flow_Augmenting_Path.

Theorem (Maximum Flow Equals Minimum Cut). Let $N(V, E)$ be a capacitated network with distinct vertices S and T. Then,

$$maxflow(N, S, T) = mincap(N, S, T).$$

6-5 MAXIMUM FLOW ALGORITHM: DINIC

The maximum flow algorithm of Dinic finds a maximum flow in a capacitated network $N(V, E)$ by repeatedly finding maximal flows in a layered subnetwork of N. The algorithm takes $|V(N)|$ search phases in contrast to the $O(|V||E|)$ phases required by the Edmonds-Karp version of the Ford and Fulkerson algorithm. We will present an enhanced version of Dinic's algorithm which uses an efficient procedure for finding maximal flows in layered networks due to Malhotra, Pramodh Kumar, and Maheshwari (1978). The algorithm has performance $O(|V|^3)$ which represents a significant improvement over the $O(|V||E|^2)$ performance of the Edmonds-Karp version of Ford and Fulkerson. (For faster, but less simple algorithms, see the references.)

Let us introduce some terminology. Let (N, S, T, f) be a conservative flow. We will say an edge y in $E(N)$ is *useful from u to v* if either y equals (u, v) and $cap(u, v) - f(u, v) > 0$ or y equals (v, u) and $f(v, u) > 0$. The *useful capacity* of an edge y useful from u to v is $cap(u, v) - f(u, v)$ if y equals (u, v) and $f(v, u)$ if y equals (v, u).

An *augmenting layered network* ALN with respect to (N, S, T, f) is then defined as follows. The set of vertices $V(ALN)$ is a subset of $V(N)$ which is partitioned into disjoint parts $L(i)$, $i = 0, \ldots, t$ where $L(0)$ equals $\{S\}$ and $L(t)$ equals $\{T\}$ and where the partition has the property that every vertex v in the partition lies on a flow augmenting path in N from S to T whose successive vertices come from successive parts $L(i)$ and $L(i + 1)$ of the partition. If u and v are vertices of ALN from parts $L(i)$ and $L(i + 1)$ respectively, the edge (u, v) or (v, u) is in $E(ALN)$ if and only if (u, v) or (v, u) is in $E(N)$ and is useful from u to v.

We refer to Figure 6-12 for an example. The parts $L(i)$ are called the *layers* or *levels* of ALN. The index t of the level that vertex T lies in is called the *length* of the layered network and is denoted by Len(ALN). We define an *advancing flow augmenting path* from S to T on a layered network as a flow augmenting path whose vertices are from successive levels of ALN. We say a flow (ALN, S, T, f) on a layered network ALN is *maximal* if there is no advancing flow augmenting path in ALN from S to T.

The key to Dinic's algorithm lies in the construction of maximal flows on successive augmenting layered subnetworks of $N(V, E)$. The procedure Dinic gives a high

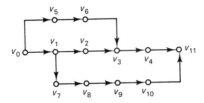

(a) Capacitated network N with unit capacities and zero flows.

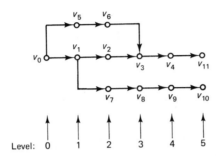

Level: 0   1   2   3   4   5

(b) First ALN before elimination of zero thruput vertices.

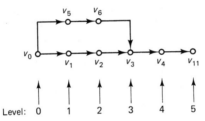

Level: 0   1   2   3   4   5

(c) First ALN after elimination of zero thruput vertices.

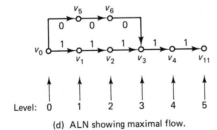

Level: 0   1   2   3   4   5

(d) ALN showing maximal flow.

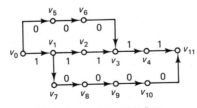

(e) N after applying ALN flow.

Figure 6-12. Example of Dinic's algorithm.

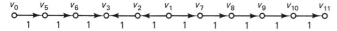

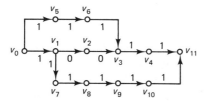

(f) Second augmenting layered network ALN'.

(g) N after applying ALN' flow.   **Figure 6-12**   (continued)

level view of this algorithm. We naturally assume that the original flow on N is identically zero, and that at each iteration the function Layered_Network(N,S,T,ALN) finds an augmenting layered network *ALN* with respect to the current conservative flow (N, S, T, f). Maximal(ALN,S,T,TRANS) constructs a maximal flow on *ALN* which it outputs as a table TRANS, which the procedure Add(TRANS,N) uses to augment the flow on N. The process is repeated as long as Layered_Network succeeds in finding an augmenting layered network with respect to the updated flows. Initial (N,TRANS) merely initializes TRANS using $N(V, E)$. See Figure 6-13 for a data flow diagram of the algorithm.

Procedure Dinic(N,S,T)

var   N: Flow Network
          ALN: Augmenting Layered Network
          TRANS: Translation Table
          S,T: 1..|V(N)|
          Layered_Network: Boolean function

Initial (N,TRANS)

while Layered_Network (N,S,T,ALN)   **do**   Maximal (ALN,S,T,TRANS)
                                                Add (TRANS,N)

End_Procedure_ Dinic

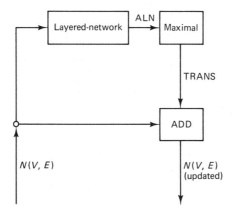

Figure 6-13.   Data flow diagram for Dinic's algorithm.

Data structures. We will describe the data structures first. Dinic uses three nonelementary types:

N: Flow Network

ALN: Augmenting Layered Network

TRANS: Translation Table

We will represent *N* and *ALN* as linear arrays. **TRANS** is an array which lets us directly access the edge representatives in *N*.

The type definition of *N* (Flow Network) is just a slightly enhanced capacitated network and follows. Its edges are represented in a shared manner on the adjacency lists of its endpoints.

type Flow Network =
    record
        NH(|V(N)|): N-Vertex
    end

  N-Vertex = **record**
        Positional-Pointer, Successor: N-Edge pointer
    end

  N-Edge = **record**
        Shared-rep: N-Shared-edge pointer
        Edge-successor,
        Edge-predecessor: N-Edge pointer
    end

  N-Shared-edge =
    record
        E(1,2): 0..|V|
        cap, flow: Nonnegative Integer
        Edge-index: 1..|E(N)|
    end

Most of the fields are either familiar or self-explanatory. The N-Shared-edge record contains a field Edge-index which gives the index in the array **TRANS** of the edge given by the 1×2 array $E(1, 2)$. Its use will be explained when we define the type of **TRANS**.

The type definition of *ALN* is designed with several objectives in mind. It must facilitate fast access to the vertices of the augmenting layered network on a given level, which we do by packaging within the representation of *ALN* a linear array L of the vertices by level. Since both the procedures Layered-Network and Maximal delete vertices and edges, we design the representation to facilitate this by including pointers in each shared edge representative in *ALN* that point back to the adjacency list entries for its endpoints, and we use doubly-linked adjacency lists for the same reason. *Lazy deletion* can be used when deleting vertices from the level lists. That is, at the point at which a vertex is to be deleted, we set its Level field to indicate a dummy level. Subsequently, when we actually have occasion to scan the level lists, we can check the Level fields of the vertices at that point, and delete a vertex from the level list if a dummy value in the entry indicates that that is appropriate. The Layered Network type definition follows.

```
type   Layered Network =
            record
                ALNH(|V(N)|): ALN-Vertex
                L(0..|V(N)| − 1): Level-Vertex
            end

       ALN-vertex =
            record
                Level: 0..|V(N)|
                Input, Output, Thruput, Flow: Nonnegative Integer
                Positional-Pointer, Successor: ALN-edge pointer
            end

       ALN-edge =
            record
                ALN-Shared-edge: Shared-ALN-edge pointer
                Edge-successor, Edge-predecessor: ALN-edge pointer
            end

       Shared-ALN-edge =
            record
                E(1,2): 0..|V|
                cap, flow: Nonnegative Integer
                Edge-index: 1..|E(N)|
                Back-ref(2): ALN-edge pointer
            end

       Level-vertex =
            record
                Positional-Pointer, Successor: Level-entry pointer
            end

       Level-entry =
            record
                Identifier: 1..|V(N)|
                Successor: Level-entry pointer
            end
```

Once again, most of the fields are either familiar or self-explanatory. The fields for ALN-vertex are defined as follows. Level is on $0..|V|$, with the value $|V|$ used to indicate a dummy level. The *input capacity* of a vertex v (denoted Input(v)) in level i of ALN is the total of the useful capacities on edges useful from level $i − 1$ of ALN to v. The *output capacity* of a vertex v is the total of the useful capacities on edges useful from v to vertices on level $i + 1$, denoted Output(v). The *thruput capacity* of v equals min(Input(v), Output(v)) and is denoted by Thruput(v). The pair of pointers in Back-ref(2) in Shared-ALN-edge point to the adjacency list entries for the endpoints of the represented edge and facilitate deletion. The Edge-index field in Shared-ALN-edge serves a similar role to the identically named field in the representation for $N(V,E)$.

The breadth first search in procedure Layered-Network uses a queue with entries of type

```
type  Queue = record
                Identifier: 1..|V(N)|
                Queue-successor: Queue pointer
              end
```

The purpose of the table **TRANS** is to store the maximal flow values determined by Maximal for *ALN* in a form that allows them to be readily applied to update the flow in $N(V, E)$. The type definition is as follows.

```
type  Translation Table =
        record
          T(|E(N)|): Table entry
        end
```

```
Table entry =
        record
          E(1,2): 1..|V|
          N-edge-entry: N-Shared-edge pointer
          Amount: Integer
          Direction: (Backward,Forward,Undefined)
        end
```

The fields have the following meanings. The array **T** has $|E(N)|$ entries since there are $|E(N)|$ edges in $N(V, E)$. Each edge in N is represented by the E field of one of the table entries. The pointer N-edge-entry allows direct access to the network representation of the edge. Recall that both N and *ALN* contain indices Edge-index which allow direct access into the array T. Layered-Network copies these indices from N to *ALN* when it constructs *ALN*. Maximal then uses these indices to store the maximal flow values on *ALN* directly in **TRANS**. Add then uses **TRANS** to directly update the flows on $N(V, E)$. The fields Amount and Direction are determined by Maximal and define the magnitude and sign of the flow changes to be made by Add to N. Prior to each operation of Maximal, the values of Amount and Direction are considered Undefined, and remain so if the edge is not involved in a flow change.

Construction of the augmenting layered network.

The algorithm for constructing an augmenting layered subnetwork of $N(V, E)$ has two phases. First, it performs a modified breadth first search which uses only useful edges to extend the search network it constructs. Then, it removes any excess vertices and edges which do not lie on flow augmenting paths from S to T. The capacities and flows of the edges in *ALN* are assigned as follows. Let (c, f) be the $(cap, flow)$ for an edge x in N which lies in *ALN* and let (c', f') denote the $(cap, flow)$ x is to be assigned in *ALN*. If x is a forward edge of *ALN*, then we set c' to $c - f$ and f' to 0 (corresponding to the useful capacity of x). If x is a backward edge of *ALN*, we set both c' and f' to f, which is again appropriate considering the useful capacity of x in this case.

Layered_Network calls several subordinate functions and procedures. We use a (lexically overloaded) procedure Reset to perform the obvious resetting functions for N and *ALN*. We use a variant of the familiar function Next(N, Head(Q),v,v-edge-ptr)

to return not only the index of the next neighbor of Head(Q) in $N(V,E)$, but also a pointer v-edge-ptr to the corresponding entry on the adjacency list of Head(Q). The procedure Useful(N,Head(Q),v,v-edge-ptr) succeeds if the referenced edge is useful from Head(Q) to v, and fails otherwise. Add (v, Level(Head(Q)) + 1) adds v to Level (Head(Q)) + 1 of *ALN*. Put(v-edge-ptr, ALN) inserts the edge referenced by v-edge-ptr in the appropriate level of *ALN*, assigns its capacity and flow correctly, and updates any affected fields in *ALN* such as Input, Output, and Thruput.

After the initial breadth first draft of *ALN* is constructed, we then proceed to remove vertices not lying on flow augmenting S to T paths in this draft of *ALN*. These vertices correspond to vertices lying at the same level as T, as well as vertices of zero Thruput at lower levels. As we remove these vertices, we may introduce additional such vertices, since the removal process may affect the Thruput of the neighbors of removed vertices. We start the process at Level(T) and proceed downward through the lower levels. The zero Thruput vertices at each level are stored on a queue *ZTQ*. An enhanced queue operation Dequeue(ZTQ,v) returns the head of the queue in v, if it exists, and fails otherwise. The procedure Remove(ALN,v) removes v from *ALN*. This is a complex operation which includes deleting edges to the neighbors of v (which are necessarily on Level(v) − 1)), updating affected Inputs, Thruputs, and the like.

```
Function Layered_Network(N,S,T,ALN)

(* Constructs an augmenting layered network ALN for (N,S,T,f),
   or fails *)

var  N: Flow Network
     ALN: Augmenting Layered Network
     S,T,v: 1..|V(N)|
     v-edge-ptr: N-Edge pointer
     M, Inc: Nonnegative Integer
     Dir: (Forward, Backward)
     Next, Empty, Dequeue, Layered_Network: Boolean function
     Head: Integer function
     Q, ZTQ: Queue pointer

Reset(N); Reset(ALN)
Create(Q); Enqueue(Q,S)
Set M to |V(G)|

(* Breadth First Advance to T *)

repeat

   while  Next(N,Head(Q),v,v-edge-ptr)

      if  Level(v) > Level(Head(Q))
             and*  Useful(N,Head(Q),v,v-edge-ptr)

      then if Level(v) = M then Add(v,Level(Head(Q)) + 1)
                                Enqueue (Q,v)
                           Put(v-edge-ptr, ALN)
```

until Empty (Dequeue(Q)) or* Level (Head(Q)) = Level (T)

(* Removal of ALN vertices not on flow augmenting paths in ALN
   from S to T *)

if not Empty (Q)

then **Set** Layered_Network to True

    Set ZTQ to L(Level(T)) − {T}

    for i = Level(T) − 1 to 0 **do**

        while Dequeue(ZTQ,v) **do** Remove(v, ALN)

        for v in L(i) **do if** Thruput(v) = 0 **then** Enqueue(ZTQ,v)

else **Set** Layered_Network to False

End_Function_Layered_Network

Construction of the maximal flow on ALN.

Maximal(ALN,S,T,TRANS) constructs a maximal flow on the layered network *ALN*, that is, a conservative flow (ALN, S, T, f') for which there are no advancing flow augmenting paths from S to T. The idea is to find a vertex v in *ALN* of minimum thruput capacity Thruput(v) and then route Thruput(v) units of flow from S to T through v. We first push Thruput(v) units of flow upwards from v towards T via useful edges and then pull Thruput(v) units through v from the direction of S. The flow can never be blocked at intermediate vertices regardless of how the routing is done because Thruput(v) is the minimum thruput capacity of any vertex currently in *ALN*, whence a blockage is impossible. If the routed flow reduces the thruput capacities of any vertices to zero, we remove them in the same manner as was done in Layered-Network. The procedure stops when *ALN* becomes disconnected. Until then, we repeatedly select successive vertices of minimum thruput and repeat the routing process.

Maximal uses a function Select(ALN,v) to return a vertex v in *ALN* (other than S or T) of minimum thruput capacity. Select fails if the minimum thruput capacity is zero, and succeeds otherwise. The procedure Reset reinitializes the translation table **TRANS**. Another procedure Route is used to move flow accumulated at one level of *ALN* to an adjacent level of *ALN*. Route(ALN,i,x) routes the accumulated flow at level i to level $i + x$, where x is restricted to be plus or minus one. Route calls Nextlevel(ALN,i+x,u,w,uw-edge-ptr) to return the next neighbor w of u on level $i + x$, and a new version of Useful to test the utility of the edge for augmentation. The adjustments made to the edge flows (by Route) are positive or negative depending on whether the edges are forward or backward in *ALN*.

Procedure Maximal(ALN,S,T,TRANS)

(* Construct maximal flow in the layered network ALN from S to T,
   which is returned in TRANS *)

```
var  ALN: Layered Network
     TRANS: Translation Table
     S,T,v: 1..|V|
     i, Lev: 1..|V(N)| − 1
     Save_Thruput: Integer
     ZTQ: Queue pointer
     Disconnected: Boolean

Reset (TRANS)
for v ∈ V(ALN) do Set Flow(v(ALN)) to 0
Set Disconnected to False

while  not Disconnected  and*  Select(ALN,v)  do

    (* Move Thruput(v) units of flow through v *)

    Set Save_Thruput to Thruput(v)
    Set Flow(v(ALN)) to Save_Thruput
    for i = Level(v) to Len(ALN) − 1 do Route(ALN,i,+1)
    Set Flow(v(ALN)) to Save_Thruput
    for i = Level(v) to 1 do Route(ALN,i,−1)

    (* Remove vertices of zero Thruput *)

    Create (ZTQ)

    for Lev = Level(T) − 1 to 1  do

        for v in L(Lev) do if Thruput(v) = 0  then  Enqueue(ZTQ,v)

        while  Dequeue(ZTQ,v)  do  Store edges at v in TRANS
                                   Remove(v,ALN)

        if Empty(L(Lev))  then  Set Disconnected to True

for v in V(ALN) do Store edges at v in TRANS

End_Procedure_Maximal

Procedure Route(ALN, i, x)

(* Route the flow accumulated at level i of ALN to level i + 1
   or level i − 1 as determined by i and x *)

var  ALN: Layered Network
     i: 1..|V(N)| − 1
     x: (+1, −1)
     u,w: 1..|V|
     uw-edge-ptr: ALN-Edge pointer
     f: Nonnegative Integer

for  u ∈ L(i)  do
```

while Flow(u) > 0 **do**

Next-level(ALN,i,u,x,w,uw-edge-ptr)

if   Useful(ALN,x,u,w,uw-edge-ptr)

then   Set f to min {Flow(u), useful capacity of uw-edge-ptr}
Change the flow on uw-edge-ptr (using f) and Thruputs
Set Flow(u) to Flow(u) − f
Set Flow(w) to Flow(w) + f

End_Procedure_ Route

Add(TRANS,N) updates the flow on $N(V,E)$ using **TRANS** to locate the edges whose flows have to be altered and to indicate the magnitude and direction of the adjustments to be made. The value in the amount field on an edge x listed in **TRANS** is added to the flow field of the corresponding edge in N, if the direction field of x in **TRANS** is forward, and is subtracted if the direction is backward.

Illustrative examples are given in Figures 6-12, 6-14, and 6-15.

Correctness and performance.   The correctness of Dinic's algorithm follows easily from a proof of its termination, just as in the case of the Ford and Fulkerson maximum flow algorithm. Therefore, we shall consider the performance of the algorithm first and its correctness afterwards. First, we will introduce some notation. Let us denote the jth augmenting layered network constructed by Dinic's algorithm from a given network N by Layered(j) and the ith level of Layered(j) by $L(i,j)$, or by $L(i)$ if j is clear from the context. If a vertex v lies in $L(i,j)$, then Level(v,j) equals i, the index

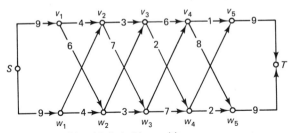

Edges labelled with capacities.
Minimum thruput vertex v_3 of thruput 8

(a) Augmenting layered network showing capacities.

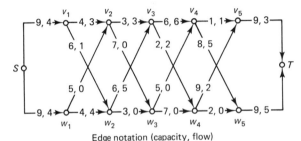

Edge notation (capacity, flow)

(b) Blind flow routing through v_3.

Figure 6-14.   Maximal flow on layered network of (a).

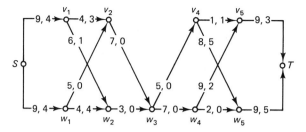

Deleted elements

v_3	Level 3
Edge	Flow
(v_2, v_3)	3
(w_2, v_3)	5
(v_3, v_4)	6
(v_3, w_4)	2

Minimum thruput vertex v_4 of thruput 3

(c) Elimination of vertices of zero thruput.

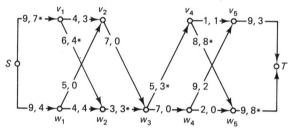

(d) Blind routing through v_4.

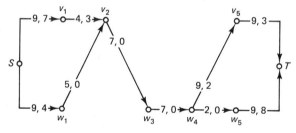

Store deleted elements

v_4	Level 4
w_2	Level 2
Edge	Flow
(w_3, v_4)	3
(v_4, v_5)	1
(v_4, w_5)	8
(v_1, w_2)	4
(w_1, w_2)	4
(w_2, w_3)	3

Minimum thrucap vertex w_5, thrucap 1

(e) Elimination of vertices of zero thruput.

Figure 6-14 (continued)

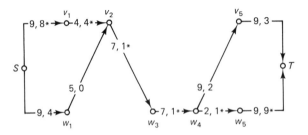

(f) Blind flow routing through w_5.

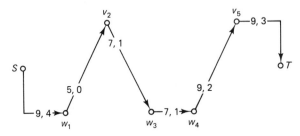

Deleted vertices

w_5, v_1

Edge	Flow
(w_4, w_5)	1
(w_5, T)	9
(S, v_1)	8
(v_1, v_2)	4

Minimum thrucap vertex w_1 of thruput 5.

(g) Elimination of zero thruput vertices w_2, w_5.

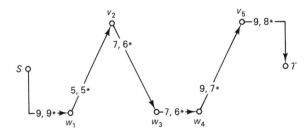

(h) Blind flow routing through w_1.

Figure 6-14 (continued)

of the level in Layered(j) that v lies in. If the parameter j is clear from the context, we use the shorthand notation Level(v). We denote the index of the level the sink T lies in, Level(T, j), by Len(Layered(j)). The performance of the algorithm relies on the fact that Len(Layered(j)) is monotonically increasing in j.

Theorem (Monotonicity of Dinic's Layered Networks). Let $N(V, E)$ be a capacitated network and let S and T be a pair of distinct vertices in N. Let $(N, S, T, 0)$ denote a trivial (zero-valued) conservative flow. If we denote the ith layered search

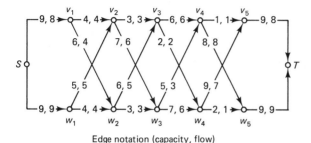

Edge notation (capacity, flow)

Figure 6-15. Reconstituted maximal layered flow network.

network constructed by Dinic's algorithm by Layered(i), starting from the trivial flow $(N, S, T, 0)$, Len(Layered(i)) is strictly monotonically increasing in i.

The proof of this theorem is as follows. Let Len(Layered($k + 1$)) be denoted by n and let P be an advancing flow augmenting path from S to T in Layered($k + 1$). Denote the successive vertices of P by w_i, $i = 0..n$. We will distinguish two cases according to whether or not every vertex in P lies in Layered(k).

Case 1: Every vertex in P lies in Layered(k)

We first establish the following claim:

Claim: Level(w_i, k) $\leq i$, for $i = 0, \ldots, n$.

The proof of the claim is by induction on i. The claim is trivial for $i = 0$. Suppose the claim is true for $i = j$. We shall prove it for $i = j + 1$. That is, we shall show that Level(w_{j+1}, k) $\leq j + 1$. Otherwise, w_{j+1} would lie in level $L(x, k)$ where $x \geq j + 2$. But, by induction, Level(w_j, k) $\leq j$. Therefore, the edge (w_j, w_{j+1}) would be from $L(y, k)$, $y \leq j$ to $L(x, k)$, $x \geq j + 2$. Therefore, (w_j, w_{j+1}) would not be in Layered(k), since by construction all the edges in a layered network are between adjacent levels. Therefore, (w_j, w_{j+1}) would not have been useful at stage k. Since (w_j, w_{j+1}) is an edge of Layered($k + 1$) by definition, it must be useful at stage $k + 1$. But, since it was not changed in stage k, it must also have been useful at stage k. Since w_j and w_{j+1} are both in Layered(k), (w_j, w_{j+1}) must lie in Layered(k), which is a contradiction. Thus, w_{j+1} cannot belong to $L(x, k)$, $x \geq j + 2$. Therefore, it must belong to $L(r, k)$, $r \leq j + 1$, as was to be shown for the claim.

Returning to the proof in case 1, we observe that it follows from the claim that Level($T, k+1$) $\geq$ Level(T, k), and so Len(Layered($k + 1$)) $\geq$ Len(Layered(k)). Furthermore, the inequality must be strict, for otherwise we can prove that P is an advancing flow augmenting path in Layered(k) after the maximal flow for Layered(k) has been established, which is a contradiction. To prove this, observe that w_i must be in $L(i, k)$. Otherwise, we can use the same argument as used in the proof of the claim to identify an edge (w_i, w_{i+1}) that skips a level in Layered(k) which would again lead to a contradiction, forcing us to include (w_i, w_{i+1}) in Layered(k) and to conclude that w_i and w_{i+1} must be in successive levels of Layered(k), showing w_i would lie in $L(i, k)$. A similar argument shows that the edges (w_i, w_{i+1}) of P must also lie in Layered(k). This proves that P is a flow augmenting path in Layered(k). But, since P is augmenting at the beginning of the phase that constructs Layered($k + 1$), P must also be augmenting at the end of the maximal flow phase that constructs Layered(k), contradicting the correctness of the maximal flow algorithm for Layered(k). It follows that Len(Layered(k)) < Len(Layered($k + 1$)), as was to be shown.

Case 2: Some vertices in P are not in Layered(k)

Let w be the first vertex in P which is not in Layered(k) and let u be predecessor of w on P. The edge (u, w) cannot be in Layered(k), since w is not in Layered(k). But, since (u, w) is augmenting in Layered($k + 1$) and is not in Layered(k), then (u, w) must also be augmenting at stage k. Since u is in Layered(k) and (u, w) is augmenting but not in Layered(k), Level(w, k) $\geq$ Level(T, k). By the same argument as in Case 1, Level($w, k+1$) $\geq$ Level(w, k), which implies Level($w, k+1$) $\geq$ Level(T, k). Consequently, Level($T, k+1$) $\geq$ Level(T, k) + 1, was to be shown. This completes the proof of the theorem.

The performance of the algorithm can be summarized as follows.

Theorem (Dinic Performance). Let $N(V, E)$ be a capacitated network and let S and T be a pair of distinct vertices in N. Then, Dinic's algorithm, enhanced by the maximal flow technique of Malhotra, Pramodh Kumar, and Maheshwari, takes $O(|V|^3)$ time.

The proof of this theorem is as follows. By the previous theorem, the algorithm constructs at most $|V(N)|$ layered networks. Thus, the performance estimate depends on the time taken by the maximal flow algorithm on the layered network. The key point is that at each flow routing phase in Maximal we use at most one edge without saturating it, so that the number of edges used is at most $O(|E(N)| + |V(N)|^2)$. Thus, the overall complexity of Dinic is $O(|V(N)|^3)$. This completes the proof of the theorem.

Theorem (Correctness of Dinic Algorithm). Let $N(V, E)$ be a capacitated network and let S and T be a pair of distinct vertices in N. Then, the flow constructed by Dinic's algorithm is a maximum flow.

The proof of this theorem is as follows. The bound of the Monotonicity theorem ensures that the algorithm terminates. Since the termination condition is that the layered search network *ALN* does not reach the sink T, just as in the case of the Ford and Fulkerson algorithm, (V(ALN), V(N) − V(ALN)) determines a cut whose forward edges are saturated and whose backward edge flows are all zero. It follows from the earlier theorems on net and maximum flow that the flow must be maximum at this point. This completes the proof of the theorem.

6-6 FLOW MODELS: MULTIPROCESSOR SCHEDULING

We can optimally schedule the execution of programs on a pair of processors using flow techniques (Stone [1977]). Let us assume that the system to be scheduled consists of separate modules which may be either executable or data modules. The execution time of an executable module depends on which processor it is scheduled to execute on; while, on the other hand, if a data module is accessed by an execution module on a different processor, then a communication cost is incurred. The objective is to assign the modules to the processors in such a way as to minimize the combined execution and communication costs.

Denote the processors by P_1 and P_2 and the modules by $M_1, \ldots, M_k$. The execution time required by an executable module M_i executing on processor P_j is denoted by T_{ij}. We assume that the modules scheduled to execute on a given processor execute sequentially. We assume further that the scheduling assignment is static. That is, each module remains with its assigned processor for the duration of the problem. In order to make an intelligent choice of module-to-processor allocation, we also assume that we know the frequency with which the different modules reference each other. If a pair of modules reside on the same processor, their intermodule communication cost is assumed to be zero. If a pair of modules M_i and M_j reside on different processors, the communication cost between them is denoted by M_{ij}. The intermodule communication costs are specified by a *communication cost graph* and the module-processor execution costs by a table. The cost of a schedule (that is, a module-to-processor assignment for every module) is the sum of the intermodule communication costs M_{ij} and the module execution costs T_{ij}. The objective is to find a schedule of minimum cost.

We model the assignment of modules to processors by a *module assignment graph* defined as follows.

(1) Add two vertices S_1 and S_2 to the communication cost graph, corresponding to the pair of processors P_1 and P_2.

(2) Add edges (M_i, S_2) of weight M_{i1} corresponding to the possible assignment of module M_i to P_1.

(3) Add edges (M_i, S_1) of weight M_{i2}, corresponding to the possible assignment of module M_i to P_2.

Figure 6-17 shows the module assignment graph for the configuration shown in Figure 6-16. With this definition of the module assignment graph, there is a 1-1 correspondence between processor-to-module assignments and (S_1, S_2) cuts. Namely, we assign the module whose vertices lie in the S_1 part of the cut to P_1, and we assign the modules in the S_2 part of the cut are to P_2. The weight of the resulting cut is equal to the cost of the assignment.

(1) The modules M_i and M_j will incur the communication cost M_{ij} if and only if M_i and M_j are on opposite sides of the cut; so the edge (M_i, M_j) is in the cut, and

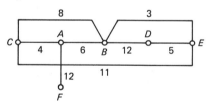

(a) Inter-module communication cost graph.

Module	P_1 cost	P_2 cost
A	5	10
B	2	M (denotes a large value)
C	4	4
D	6	3
E	5	2
F	M	4

(b) Module-processor execution cost table.

Figure 6-16. Communication and execution costs for modules.

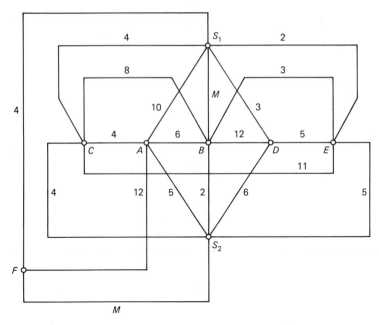

Figure 6-17. Module assignment graph.

(2) Module M_i incurs the execution cost T_{i1} if and only if (M_i, S_2) is an edge of the cut (similarly for T_{i2}).

Thus, the cost of the cut equals the cost of the module-to-processor assignment. It follows from the theory of network flow that the optimal module-to-processor assignment is obtained by finding the maximum value flow from S_1 to S_2. Figure 6-18 shows an optimal solution for the example of Figure 6-17. Figure 6-19 illustrates a suboptimal assignment.

6-7 CONNECTIVITY ALGORITHMS

We can compute the vertex and edge connectivity of a graph by modelling connectivity as a network flow problem, and then applying maximum flow techniques. There are a variety of connectivity problems to consider. The *pairwise connectivity* problem asks for the vertex (or edge) connectivity between a given pair of vertices in a graph. The *pairwise disjoint paths* problem requires finding a maximum cardinality set of vertex (or edge) disjoint paths between a given pair of vertices. The *pairwise disconnecting sets* problem seeks a minimum size vertex (or edge) disconnecting set between a given pair of vertices; while the *graph connectivity* problem seeks the vertex (or edge) connectivity of a graph. We begin by considering the vertex variations of these problems.

Problems of vertex connectivity. Let $G(V, E)$ be a graph. We shall construct a flow network called the *associated flow network* G' for G as follows:

(1) For each vertex v in $V(G)$, create a pair of vertices v' and v'' and an edge (v', v'') in $V(G')$;

(2) For each edge (u, v) in $E(G)$, create a pair of edges (u'', v') and (v'', u') in $E(G')$;

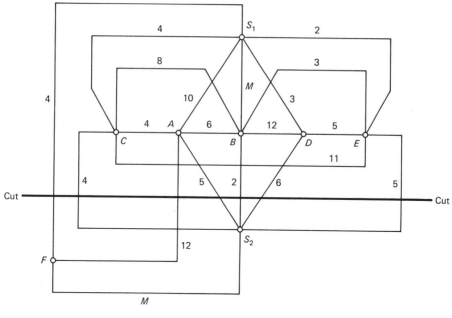

Cut: $X = \{S_1, A, B, C, D, E\}$, $X^c = \{S_2, F\}$

(a) Optimal assignment with minimum cut.

Processor	Task				
P1	A	B	C	D	E
P2	F				
Execution Cost	26				
Communication Cost	12				
Assignment Cost	38				

(b) Assignment cost of optimal assignment.

Figure 6-18. Optimal assignment.

(3) Assign a unit capacity to each edge in $E(G')$ of the form (v', v''), and some large capacity M (that plays the role of plus infinity) to each edge created in (2).

This construction allows us to translate problems about paths in G into problems about flows in G'. The construction is illustrated in Figure 6-20.

The associated flow network satisfies the following important property.

Claim: Let u'' and v' be vertices in G' and let F be a set of paths from u'' to v', each of unit capacity, which realize a flow from u'' to v'. Then, the paths in F are vertex disjoint, except for their endpoints u'' and v'.

The claim follows because any internal vertex on any of the paths in F has either a unique incoming edge of unit capacity or a unique outgoing edge of unit capacity. Therefore, at most one unit capacity path can pass through any internal vertex of the paths in F.

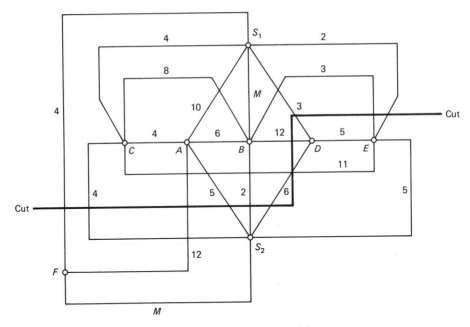

(a) Suboptimal assignment with non-minimum cut.

Processor	Task		
P1	A	B	C
P2	D	E	F
Execution Cost	20		
Communication Cost	38		
Assignment Cost	58		

(b) Assignment cost for suboptimal assignment.

Figure 6-19. Suboptimal assignment.

We can map a set of u'' to v' flow paths in G' to a set of u to v paths in G by the simple expedient of contracting flow edges of the form (x', x'') to a vertex x and replacing flow edges of the form (x'', y') by an edge (x, y).

Conversely, we can map a u to v path in G to a u'' to v' flow path in G' as indicated by the following example: The path u-a-b-v in G becomes u''-a'-a''-b'-b''-v' in G'.

It follows readily from these observations that

(1) A maximum cardinality set of vertex disjoint paths between a given pair of vertices u and v in a graph G corresponds to a set of flow paths realizing a maximum flow between u'' and v' in G';

(2) The pairwise connectivity $VC(G, u, v)$ between a pair of vertices u and v in G equals the maximum flow value between the vertices u'' and v' in G'; and

(3) A minimum size vertex disconnecting set between a pair of vertices u and v in G corresponds to a minimum capacity $u'' - v'$ cut in G'.

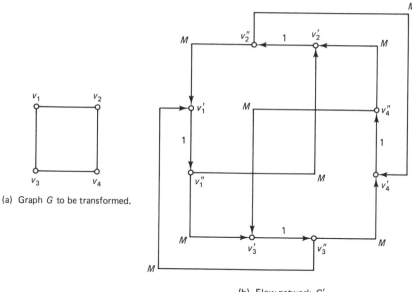

(a) Graph G to be transformed.

(b) Flow network G'.

Figure 6-20. Associated flow network.

Statement (3) follows by observing that a finite capacity cut in G' must contain only edges of the form (x', x''). Therefore, the vertices x in G that correspond to the edges (x', x'') of a cut in G' comprise a vertex disconnecting set in G of cardinality equal to the capacity of the cut.

The correspondences indicated by these statements allow us to calculate connectivities on G using flow algorithms on G', as intended. We refer to Figure 6-20 for an illustration. $VC(G, v_1, v_4)$ equals 2, so there are a maximum of two vertex disjoint paths between v_1 and v_4. The vertices $\{v_2, v_3\}$ constitute a minimum size vertex disconnecting set. As expected, the maximum value flow from v_1'' to v_4' in G' is also 2. The paths in G' realizing the maximum flow are v_1''-v_2'-v_2''-v_4' and v_1''-v_3'-v_3''-v_4'. If we contract edges on the flow paths of the form (x', x''), we obtain a maximum set of vertex disjoint paths in G, namely: v_1-v_2-v_4 and v_1-v_3-v_4. A minimum capacity cut between v_1'' and v_4' in G' is given by $(\{v_1'', v_3', v_2'\}, \{v_1', v_2'', v_3'', v_4', v_4''\})$. The edges of the cut are (v_2', v_2'') and (v_3', v_3''). Therefore, $\{v_2, v_3\}$ is the corresponding minimum size vertex disconnecting set.

Vertex connectivity of a graph. The usual formula for the vertex connectivity of a graph $G(V, E)$ in terms of its pairwise vertex connectivities is

$$VC(G) = \min_{(u, v) \text{ not in } E(G)} \{VC(G, u, v)\}.$$

The following procedure improves on this slightly. For convenience, we assume that if G is not a complete graph; then its vertices are ordered so that v_1 is not adjacent to some vertex v. We invoke functions $VC(G, u, v)$ to calculate the pairwise vertex connectivities.

```
Function VC(G)

(* Returns in VC the vertex connectivity of G *)

var  G: Graph
     i, j: 1..|V(G)|
     VC: Integer function

Set VC to |V(G)| − 1
if G is complete then return

for i = 1 to |V(G)| do

   for j = i + 1 to |V(G)| do

      if     i and j not adjacent
      then   Set VC to min{VC,VC(G,i,j)}

   if  i > VC  then  return

End_Function_VC
```

Theorem (Correctness of VC).   Let $G(V, E)$ be a graph. Then, the function $VC(G)$ correctly calculates the vertex connectivity of G.

The proof is as follows. If G is complete, the procedure is trivially correct. Otherwise, if G is not complete, the first calculation of $VC(G, i, j)$ with i and j not adjacent already forces VC to at most $|V| - 2$. Therefore, the procedure must terminate with an explicit **return** at least by the time the outer loop is executed with $i = |V| - 1$. Consequently, upon termination,

$$VC < i.$$

Since $VC(G)$ is the minimum of the nonadjacent pairwise connectivities, VC is always at least as large as $VC(G)$; therefore

$$VC(G) + 1 \le VC + 1 \le i.$$

Since $VC(G)$ is the cardinality of a minimum cardinality vertex disconnecting set S for G and since i exceeds VC on termination, at least one vertex j in $1 . . i$ cannot lie in S. Since j is not in S, j is separated from some other vertex v by the disconnecting set S. Refer to Figure 6-21. Necessarily, $VC(G, j, v) \le |S|$. Since S is a disconnecting set of minimum cardinality, $VC(G, j, v)$ must equal $|S|$, that is, $VC(G)$. Therefore, VC will receive its correct value when the algorithm calculates $VC(G, j, v)$. This completes the proof.

Problems of edge connectivity.   We can calculate $EC(G)$ by finding a maximum flow on a network G' obtained from G by replacing each undirected edge $\{u, v\}$ in G by a pair of directed edges (u, v) and (v, u) in G' and assigning a unit capacity to each edge of G'. The pairwise edge connectivity, pairwise edge disjoint path, and pairwise edge disconnecting set problems can then be solved using flow methods. The flow bearing paths of a flow on G' correspond directly to a set of edge disjoint

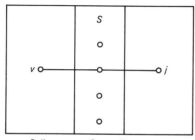

S disconnects G as well as v and j

Figure 6-21. VC(G) calculation.

paths in G. The edges of a minimum cut between a pair of vertices in G' determine a minimum cardinality edge disconnecting set in G between those vertices. We can calculate $EC(G)$ by selecting an arbitrary vertex v in G and calculating the minimum of $EC(G, u, v)$ over all distinct vertices u in G, and the given v (why?).

Probabilistic algorithm for VC(G). A probabilistic algorithm for a graphical invariant calculates the exact value of the invariant with a high degree of probability, though not with certitude. We can calculate VC(G) probabilistically by observing that $VC(G)$ equals $VC(G, u, v)$ for any pair of vertices u and v in G which are separated by a minimum cardinality vertex disconnecting set. The idea is as follows. Let $\{v_1, \ldots, v_k\}$ be a random set of k vertices of G. If any of these vertices is not in some minimum vertex disconnecting set S, say vertex v_j, then $VC(G, v_j, v)$ equals $VC(G)$ for some vertex v which is separated from v_j by S. We can estimate the probability of having at least one such vertex v_j in a set of k random such vertices as follows. The probability that a given random vertex v lies in a particular minimum vertex disconnecting set S of cardinality $VC(G)$ is $VC(G)/|V|$. Therefore, the probability that at least one of k random vertices lies outside S is at least $1 - (VC(G)/|V|)^k$. For example, if $|V|$ is 100 and $VC(G)$ is 10, then the probability that at least one of three randomly selected vertices lies outside S is at least 99.9%. The probabilistic procedure follows.

```
Function Random_VC(G,k)

(* Returns the value of VC(G) in Random_VC with probability
   1 − (VC(G)/|V(G)|)ᵏ of being correct *)

var  G: Graph
     k: Integer constant
     i, j, R(k): 1..|V(G)|
     Random_VC: Integer function

Set Random_VC to |V(G)| − 1

if G is complete then   return

Select k distinct random vertices {R(1),...,R(k)} from V(G)

for i = 1 to k do

  for j = 1 to |V(G)| do
```

if   R(i) and j are distinct and nonadjacent

then   Set Random_VC to min{ Random_VC, VC(G,R(i),j) }

End_function_Random_VC

6-8 PARTIAL PERMUTATION ROUTING ON A HYPERCUBE

There are a variety of routing problems of practical interest not covered by the classical theory we have described so far. For example, consider the problem of routing on the class of networks called hypercubes, a type of network used to interconnect parallel computers.

A *hypercube* $H(V, E)$ is an *n*-dimensional cube with 2^n vertices, each with an *n*-bit address, and containing an edge between any pair of vertices whose addresses differ in a single bit. Each vertex has degree *n* and the hypercube has diameter *n*, both logarithmic in the order $|V(H)|$ of the hypercube. We can visualize processors with addresses differing in exactly the *i*-th bit as geometrically adjacent along the ith dimension of the hypercube. A three-dimensional hypercube is shown in Figure 6-22, while Figure 6-23 shows part of a four-dimensional hypercube. There is a processor at each vertex, which communicates with the other processors by messages routed via the hypercube network. Each message contains the address of its target processor, and it takes unit time to transmit a message along an edge.

A basic routing problem for hypercubes is to concurrently transmit messages from one subset of its processors to another subset of its processors, so-called *partial permutation routing*. Each message contains a distinct target address, and the objective is to route the messages to their destinations as quickly as possible. Since only one message can be transmitted along an edge at a time, messages may have to be queued at processors lying on their route if there is contention for access to an edge.

We will describe two (distributed parallel) algorithms for implementing partial permutation, a greedy algorithm whose worst case performance is poor, and a randomized version of the greedy algorithm with optimal expected time performance.

Greedy partial permutation algorithm.   Permutation routing has an obvious greedy algorithm. We denote a message by *m*, the address of the destination of *m* by Target(m), the address of a vertex processor *v* by Address(v), and the index of the leftmost bit of Target(m) that differs from the corresponding bit in Address(v) by Leftmost(v,m). We can permute the messages by simply moving each message *m* lying at a

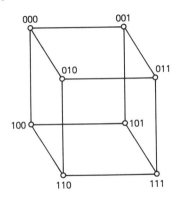

Figure 6-22.   Three-cube on eight vertices.

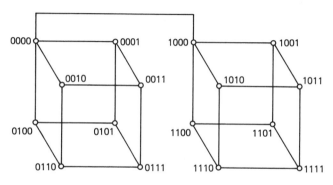

Figure 6-23. Four-cube with a single fourth dimensional connection shown.

vertex v to that neighbor of v whose address matches Target(m) in one more bit than Address(v) does. Each such step brings each message one bit closer to its destination, and along a shortest path of length at most n, though resource contention by competing messages may delay the transmission.

The following high-level procedure describes the algorithm. Messages routed through a processor v are queued on a Message_queue(v) until transmitted to the next vertex on their route. A utility Dequeue(Message_queue(v),m) returns the head of the queue in m, or fails if the queue is empty; while Enqueue(m,Message_queue(v)) queues m on the message queue at v. We use a command Send(m,w) to indicate transmission of m to w. If w is busy, m is delayed at v on a transmission queue which is managed in an interrupt-driven fashion, and is otherwise transparent. A primitive Receive(m) returns a transmitted message in m, or fails if there is no incoming message. The messages reside initially at their home locations. The same procedure is executed at each processor. The data types are only suggestive.

Procedure Greedy_Routing(v)

(* Partial permutation routing on hypercube *)

var v, w: Hypercube vertex
     m, m': message
     Dequeue, Receive: Boolean function

repeat

    if   Dequeue(Message_queue(v),m)
    then **Set** w to next neighbor of v such that Address(w) differs
           from Address(v) in bit Leftmost(v,m)
           Send(m,w)

    if   Receive (m')
    then Enqueue (m', Message_queue(v))

until   partial permutation is completed

End_Procedure_ Greedy_Routing

The problem with the greedy algorithm is that messages can be greatly delayed due to access contention at vertices along their routes, as illustrated by the following example. Suppose H is a ten-dimensional hypercube, with 1,024 vertex processors with addresses $(a_0 \ldots a_9)$. Suppose 2^5 messages have home addresses at $(0, 0, 0, 0, 0, a_5, a_6, a_7, a_8, a_9)$, and target addresses at $(a_9, a_8, a_7, a_6, a_5, 0, 0, 0, 0, 0)$. Thus, each message is targeted for a processor whose address is the reverse of its home address. The greedy algorithm forces every message through the processor at $(0, 0, 0, 0, 0, 0, 0, 0, 0, 0)$ at step 5, causing a bottleneck. This leads to a delay of $O(|V(H)|^{.5})$, which is substantially greater than the theoretical logarithmic lower bound of n.

Valiant-Brebner random routing algorithm. A randomized version of the greedy algorithm attains the $O(n)$ lower bound on transmission with probability arbitrarily close to 1 (as a function of n). The idea is to randomize the initial distribution of messages, which spreads the risk of the kind of source-to-target mapping that produces a bottleneck. Thereafter, we route the messages to their destinations using the previous deterministic greedy procedure. The technique is as follows.

(1) Each processor containing a message m to be routed, randomly generates an intermediate destination address $r(m)$ for m.

(2) The greedy algorithm is used to route each message to its random intermediate target. Until a message m reaches $r(m)$, it has priority access to network resources over messages that have already reached their randomized point of departure.

(3) Once a message m arrives at $r(m)$, we route m to its original destination Target(m) using the deterministic greedy method.

The performance of the algorithm is summarized in the following theorem.

Theorem (Valiant-Brebner Hypercube Routing). Let $H(V, E)$ be an n-dimensional hypercube and let p be any partial permutation on $V(H)$. Then, the probability that Valiant-Brebner random routing takes more than $8n$ steps is less than $O(0.74^n)$.

We refer to Valiant and Brebner (1981) for the nontrivial proof. One can show that using this method, the chance that more than $i(\log n)$ processors will simultaneously try to transmit a message through a given processor decreases exponentially with i. This makes serious bottlenecks rare. Nonetheless, a subset of processors can become embroiled in a deadlock, or the performance may happen to be poor. In each case, we can restart the algorithm with a statistically high chance of improvement on the next attempt.

REFERENCES AND FURTHER READING

For a discussion of theoretical results on connectivity, see Harary (1971) and Behzad, et al. (1979). Boesch (1976) and (1982) describe network applications. Even (1979) gives a very thorough treatment both of flow algorithms, their applications to connectivity and planarity algorithms; it was the basis of part of our discussion of how to compute connectivity using flows and for Dinic's algorithm. See Ford and Fulkerson (1962) for the fundamentals of network flow theory and its variations. Lawler (1976) includes a good discussion of multicommodity flows. The enhancement of Dinic's algorithm presented uses a technique of Malhotra, Pramodh Kumar, and Maheshwari (1978). Edmonds and Karp (1972) gives an improved version of the Ford and Fulkerson maximum flow algorithm. For a detailed discussion of other flow maximization techniques, such as the conceptually more complicated but more efficient wave method, see Tarjan (1983) and Sleator and Tarjan (1983) where an $O(|E| \log |V|)$ method for finding maximal flows is described. The interesting Byzantine routing problem is described in Dolev (1982). Stone (1977) applies flow techniques to scheduling systems of processors. Becker, et al. (1982) give the probabilistic modification to connectivity calculation. The probabilistic routing algorithm for hypercube routing is analyzed in Valiant and Brebner (1981).

EXERCISES

1. Prove that if v is an articulation point of $G(V,E)$, then v is not an articulation point of G'.
2. Construct a graph G with $VC(G) = 2$, $EC(G) = 3$, and $\min(G) = 4$. In general, try to construct a graph with $VC(G) = n - 2$, $EC(G) = n - 1$, and $\min(G) = n$.
3. Show the definitions of vertex and edge connectivity given in this chapter are equivalent to those given in Chapter 1. In particular, show that the minimum value of $VC(G, u, v)$ over all pairs of vertices u and v equals the minimum value of $VC(G, u, v)$ over all pairs of nonadjacent vertices u and v.
4. Prove the following special case of the Connectivity Lower Bound Theorem: If $\min(G) \geq |V|/2$, then $EC(G) = \min(G)$. Does an analogous result hold for $VC(G)$?
5. Prove a graph $G(V,E)$ of vertex connectivity k satisfies that $|E| \geq k|V|/2$.
6. Prove that $VC(G) = EC(G)$ if $G(V,E)$ is a cubic graph.
7. Prove that the Petersen graph is the smallest three-connected three-regular graph.
8. Prove that the complement of a disconnected graph is spanned by a complete bipartite graph.
9. Prove that if a connected graph $G(V,E)$ is r-regular and contains an articulation point, $EC(G)$ is bounded by $r/2$.
10. Does the Flow_Augmenting_Path function used for the Ford-Fulkerson maximum flow algorithm need predecessor pointers separate from the search stack if depth first search is used? What graphical object does depth first search naturally maintain? What graphical object does breadth first search naturally maintain? Can each search discipline maintain the other's natural graphical object conveniently?
11. Adapt the Ford-Fulkerson algorithm to the case where both the vertices and the edges of the network have capacities. *Hint:* Remodel the input network appropriately so that it has the format expected by the Ford-Fulkerson algorithm.
12. Let $G(V,E)$ be a digraph containing a unique vertex x of indegree 0 and a unique vertex y of outdegree 0. Try to design an algorithm to find a minimum cardinality cutset between x and y.
13. Suppose that in a network $N(V,E)$, there is not only a capacity associated with each edge, but also a cost, representing the unit cost of transmitting flow through that edge. Show how to model the minimum cost, maximum flow problem on such a network using linear programming.
14. Implement a flow-based algorithm for finding the edge connectivity of a graph.
15. Implement a flow-based algorithm for finding the vertex connectivity of a graph.
16. Implement the probabilistic algorithm for finding the vertex connectivity of a graph, and compare its performance with the corresponding deterministic algorithm for a reasonable population of test graphs.
17. Suppose every edge of a graph has the same fixed probability of failure. Use simulation to obtain an estimate that a given such graph is disconnected, due to some combination of edge

failures. For fixed order, and for a given number of edges, can you determine how to allocate the edges among the vertices so the resulting network has minimum probability of failure? Consider, as an elementary example, the case where $|E| = |V| + 1$. Obtain a theoretical estimate of the probability of failure in the case of such graphs, and compare this with an estimate derived from simulation.

18. Implement the path extraction algorithm for flow networks.

19. Simulate the probabilistic routing algorithm on a hypercube.

20. Write a program that simulates the operation of a "Byzantine" network, and that uses reliable routing methods to route messages through the network. Verify that routing can be done reliably even in the presence of simulated unreliable processors.

21. Give an example of a maximal flow on a layered network which contains an ordinary flow augmenting path, despite the absence of any advancing flow augmenting paths.

22. Is a hypercube bipartite?

7

Graph Coloring

7-1 BASIC CONCEPTS

A *k-coloring* of a graph G is a mapping of V(G) onto the integers $1 .. k$ such that adjacent vertices map into different integers. A *k-coloring* partitions $V(G)$ into k disjoint subsets such that vertices from different subsets have different colors. Of course, it can happen that a pair of different *k-colorings* may nonetheless partition $V(G)$ into the same parts, in which case, the colorings are in a sense equivalent. Given a *k-coloring*, it is customary to call the integer a vertex maps into its *color*. A graph G is *k-colorable* if it has a *k-coloring*. The smallest integer k for which G is k-colorable is called the *chromatic number* or *vertex chromatic number* (VCHR) of G. A graph whose chromatic number is k is called a *k-chromatic graph*. Figure 7-1 shows a four-chromatic graph and Figure 7-2 shows the smallest triangle-free four-chromatic graph, the Grotzsch graph. A mapping from $E(G)$ onto the set of integers $1 .. k$ such that adjacent edges map into different integers is called an *edge coloring* of G. The cardinality of a minimum cardinality edge coloring of G is called the *edge chromatic number* (ECHR) of G. There are extensive and difficult theoretical results about coloring. We will mention a few.

Theorem (Upper Bounds). If a connected graph $G(V, E)$ is neither an odd cycle nor a complete graph, then the chromatic number of G satisfies

$$\text{VCHR}(G) \leq \max(G) \qquad \text{(Brooks' Inequality)}.$$

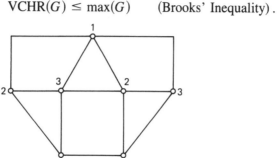

Figure 7-1. A four-chromatic graph.

223

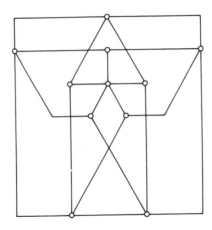

Figure 7-2. Smallest triangle-free four-chromatic graph.

For any nontrivial graph G, the edge chromatic number ECHR of G satisfies

$$\text{ECHR}(G) \leq 1 + \max(G) \qquad \text{(Vizing's Inequality)}.$$

Obviously, $\max(G)$ is a lower bound for $\text{ECHR}(G)$. Both Brooks' and Vizing's Inequalities are nontrivial to prove. However, a weakened version of Brooks' Inequality,

$$\text{VCHR}(G) \leq 1 + \max(G)$$

is easy to prove, and is realized by the following procedure. As usual, Next(G, v) returns a next vertex in G or fails:

> **while** Next(G,v) **do Set** color(v) to a color not used by the neighbors of v

The body of the **while** loop always succeeds because the neighbors of any vertex exhaust at most $\max(G)$ colors. Therefore, if we allow $\max(G) + 1$ colors, we can always color v.

The following theorem of Koenig is well known.

Theorem (Bipartite Edge Chromatic Number). If $G(V, E)$ is bipartite, $\text{ECHR}(G) = \max(G)$.

The proof is by induction on the number of edges in G. Fix the number of vertices in G, and suppose G equals $G' \cup (u, v)$ where G' has n edges and maximum degree m. If u or v is of degree m in G', $\max(G)$ is $m + 1$; hence we can color (u, v) with $m + 1$, and the induction follows. Otherwise, both u and v have degree in G' at most $m - 1$. If neither u nor v are incident with an edge colored with some color k, we can color (u, v) with k, using at most m colors for G, and again the induction follows.

Otherwise, there is some color k not incident with u and some color j not incident with v. Consider the set of vertices reachable from u by paths in G' containing only edges colored k or j. Let S be the induced subgraph on these vertices, and containing only edges colored k or j. Since G' is bipartite, any path from u to v has odd length. Therefore, any path from u to v in S must begin and end with the same color. Therefore, v cannot be in S since any path from u to v in S would begin and end with a j-colored edge which could not be incident with v. Therefore, if we interchange the colors of all the edges in S, changing j to k and k to j, we still maintain a valid edge coloring of G'. Since the edges incident with v are unaffected by this change, no edge incident with u or v is colored k in the new coloring. This corresponds to the previous case, and so completes the proof by induction.

The most well-known result in graph theory is the famous

Theorem (Four-Color Theorem).   If G is a planar graph,

$$\text{VCHR}(G) \leq 4.$$

The proof of the Four-Color Theorem is enormously complicated. The weaker result

$$\text{VCHR}(G) \leq 5$$

is much easier to prove. We will later describe an algorithm for five-coloring a planar graph based on the proof of this Five-Color Theorem.

The Four-Color Theorem relates the planarity and chromatic number of a graph. This can be generalized to more complex surfaces than planes. Indeed, there is a simple relationship between the embeddability of a graph on a manifold, the maximum chromatic number of the graph, and the so-called topological genus of the manifold. Roughly speaking, a *surface of genus N* is a sphere to which N "handles" have been added. Thus, a sphere itself has genus zero, while a torus, which can be visualized as a sphere with one handle attached to its surface, has genus 1. The *chromatic number of a surface S of genus N* is the maximum chromatic number of all graphs that can be embedded on S. In this terminology, the Four-Color Theorem asserts that the sphere (which is equivalent for embedding purposes to the plane) has chromatic number 4. The following theorem is a vast generalization of this.

Theorem (Heawood Map-Coloring Theorem).   For every positive integer N, the chromatic number of a surface of genus N is the greatest integer in

$$(7 + (1 + 48N)^{(1/2)})/2.$$

The Heawood formula (correctly) returns 4 for the chromatic number of a sphere, a surface of genus 0. It returns 7 as the chromatic number of a torus, a surface of genus 1. Observe that $K(7)$ which has VCHR $= 7$, can be embedded on the torus.

7-2 MODELS: CONSTRAINED SCHEDULING AND ZERO-KNOWLEDGE PASSWORDS

Constrained scheduling.   Colorings can be used to model scheduling activities with overlapping resource requirements. Let $T_1, \ldots, T_n$ be a set of tasks, each taking a unit time. Suppose some of the tasks cannot be scheduled concurrently because they use resources that are unshareable. We can summarize such concurrency constraints by a *conflict matrix* **C** where $\mathbf{C(i, j)}$ is 0 if tasks T_i and T_j require no common resource, and is 1 otherwise. We can interpret the conflict matrix as a *conflict graph* where each vertex corresponds to a task and there is an edge between a pair of vertices if the corresponding tasks cannot be scheduled concurrently. A k-coloring of the conflict graph partitions its vertices into k differently colored sets of tasks. If we schedule all the tasks numbered (colored) i for time period i, all the tasks will be finished by time $k + 1$. The length of a shortest possible schedule equals the chromatic number of the conflict graph. Figure 7-3 shows an example.

Zero-knowledge passwords.   The chromatic number of a graph is not easy to compute. Indeed, it is an example of what we will later call an NP-Complete problem. NP-Complete problems do not seem to be solvable by algorithms with polynomial performance bounds. Generally, this is a disadvantage. But, curiously, applications have recently been found that rely precisely on the difficulty of solving such problems.

A common security problem is to restrict access to a system to a list of valid users. We can use graph colorings to accomplish this in an extremely reliable manner. The idea is for each user to randomly generate a three-colorable graph. We use the graph as the user identification number. To access the system, a user provides a graph-id, and proves ownership of the graph by three-coloring it. The graph-id is unforgeable because, even if someone else gets a copy of the graph, it is useless without its coloring, since it is computationally intractable to find a three-coloring of a given graph. Thus, as long as the coloring is kept secure, the graph establishes the user's identity in a way that cannot be forged. Furthermore, we can actually show we can three-color a graph without even revealing the coloring by a probabilistic procedure we will describe. This allows us to use the same id repeatedly, since the coloring is never revealed. It also further enhances the security of the id since not even the system needs to know the coloring.

This is an example of a public-key cryptosystem. In such a system, part of the id for a valid user, in this case the graph, can be public. Only the coloring part has to be kept secret and that, unlike an ordinary password, need not be shared even with the system being accessed.

The following points remain to be considered:

(1) How to construct a random three-colorable graph, and

(2) How to prove one knows how to color a graph without revealing the coloring.

While it is difficult to three-color a given graph, it is easy to construct a random three-color graph inductively. Let $G(V, E)$ be a random three-color graph constructed so far. To extend G, we create a new vertex v, randomly color v with one of the three colors, and then add edges between v and a random subset of the vertices in G with different colors than v. We repeat the process until the order of G is sufficiently large. Though, by construction, we can three-color the final graph, a computational barrier prevents anyone else from duplicating this feat.

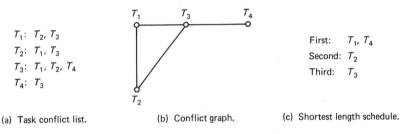

T_1: T_2, T_3		First: T_1, T_4
T_2: T_1, T_3		Second: T_2
T_3: T_1, T_2, T_4		Third: T_3
T_4: T_3		

(a) Task conflict list.   (b) Conflict graph.   (c) Shortest length schedule.

Figure 7-3.   Scheduling example.

We can convince an observer we can three-color a given graph without revealing its coloring as follows. For simplicity, we use some physical imagery. We lay the graph on a table with its vertices covered. The observer picks a random pair of adjacent vertices. We expose their colors, revealing their difference. We then secretly and randomly permute the colors, changing all vertices with color i ($i = 1, 2, 3$) to color $p(i)$, where p is a randomly chosen permutation of 1, 2, and 3. (Only a permutation of the colors is allowed. We are not allowed to change the coloring substantively.) We repeat the process until the observer is satisfied we can three-color the graph.

Of course, if we did not actually have a three-coloring of the graph, at each step of the above procedure there would be a fixed chance r the observer would select for examination a pair of vertices that were incorrectly colored. The chance we could deceive the observer in n successive trials would be at most $(1 - r)^n$, which rapidly approaches zero. Thus, after some sufficiently large number of tests (the actual number depends on r, which in turn depends on $|V(G)|$ and $|E(G)|$) the chance of deception can be made arbitrarily small. Furthermore, because we randomly permute the colors, the observer cannot combine the results of one test with that of another to deduce the whole coloring, or indeed any part of it. Since the permutation effectively erases the observer's memory, he essentially knows only for each test whether the outcome was successful or not.

7-3 BACKTRACK ALGORITHM

Finding all the valid M-colorings of a graph $G(V, E)$ is a good illustration of backtracking. We will use a vector **VCR($|V|$)** to store the colors of the vertices. At each stage of the backtrack search, **VCR(k)** equals the color currently assigned to vertex k. We take **VCR** to be initially zero, and expand the coloring one vertex at a time, repeatedly extending the partial coloring represented by **VCR(1..k − 1)** to the larger coloring **VCR(1..k)** by assigning to vertex k the next color not conflicting with its previously colored neighbors (which have indices on **1..k − 1**).

Each attempted extension of the coloring has four possible outcomes. If the extension succeeds, the procedure advances to the next vertex $k + 1$. If the extension fails, the procedure restores vertex k to an uncolored state and backs up to consider a new coloring for the preceding vertex $k − 1$. If the extension fails at the very first vertex, the procedure exits. If the extension succeeds at the last vertex, the procedure displays the complete coloring. The process continues until every valid coloring is found and displayed.

It is instructive to visualize the process in terms of a search tree. There is one level in the search tree for each vertex in the graph. The levels are indexed from 1 to $|V|$, and there is an additional (dummy) level zero. There is an edge in the tree for each possible choice of color for an edge of the graph, each tree edge corresponding to a choice of color for the lower level vertex the edge ends at. Figure 7-4 shows the tree of two-colorings for the graph of five vertices in Figure 7-5. Observe that if a connected graph has chromatic number equal to two, then all of these colorings are equivalent, in the sense that the partitions of the vertices they induce are all the same.

Each path from the root of the tree to an endpoint corresponds to an (a priori) possible coloring of the graph, though not necessarily a valid coloring. If we ignore the constraints imposed by adjacency considerations, there are $|V|^M$ ways to color a $|V|$ ver-

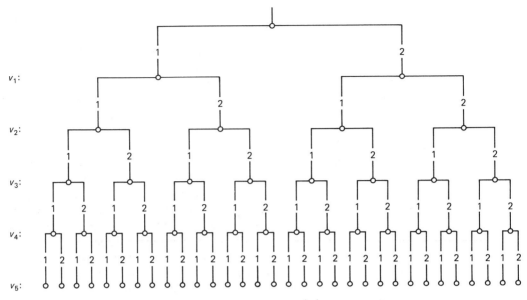

Figure 7-4. Tree of colorings: $|V| = 5$, $M = 2$.

tex graph with at most M colors, corresponding to $|V|^M$ paths from the root to the endpoints of the tree. A path from the root to a vertex at level i corresponds to a (partial) coloring of the first i vertices of a graph. The subtree rooted at the vertex at level i contains all possible continuations of the partial coloring defined by the tree path up to that vertex.

We could find the valid colorings of a graph by traversing each of its search tree paths completely, not testing the validity of the corresponding coloring until the end of the path was reached. Of course, in an intelligent search, we would terminate a particular path as soon as we recognized that a partial color assignment was invalid.

Figure 7-6 shows the tree of attempted two-colorings for the graph of Figure 7-5, with the levels arranged in the order v_1, v_2, v_3, v_4, v_5. At each tree vertex, the left (right) edge corresponds to a choice of color 1 (2) for the vertex corresponding to the lower endpoint of the edge. In this case the graph has no valid two-coloring. For example, if v_1 and v_2 are 1-colored, v_3 cannot also be 1-colored. We indicate this by a dash (-) before the corresponding edge in the search tree.

We can also order the vertices differently. This may affect the efficiency of an intelligent search. For example, in Figure 7-7 we have ordered the vertices v_1, v_4, v_5, v_2, v_3. Under this order, much more of the search tree is ignored. If we define the cost of a

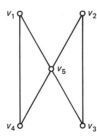

Figure 7-5. Graph for backtracking example.

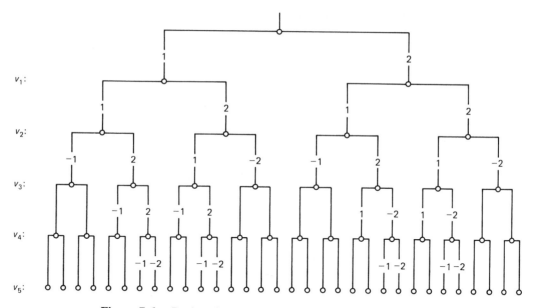

Figure 7-6. Backtrack tree with $M = 2$ ("–" denotes a deadend).

search algorithm as the number of tree edges it explores, then the traversal order of Figure 7-6 has cost 30, while the traversal in Figure 7-7 has cost 10.

In summation, backtracking allows us to traverse a search tree, while pruning unnecessary subtrees from the search. In tree terminology, a partial coloring **VCR(1 . . i)** corresponds to a search path through the tree down to a tree vertex at level i. We can extend the search path from its end vertex at level i to any of M possible successors. We can rephrase our previous rules using this terminology as follows.

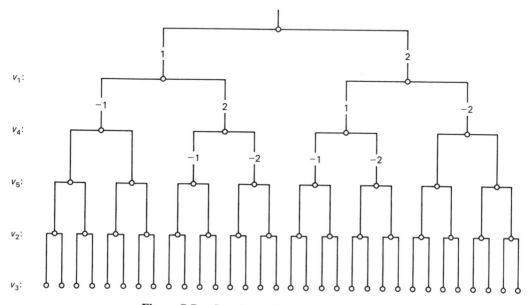

Figure 7-7. Search tree for different scan order.

(1) *Advance:* If a valid extension of the search path is available, the search path advances to the next level in the tree.

(2) *Back Up:* If no valid extension is available, the search path backs up to its previous vertex, and the search process continues from there.

(3) *Exit:* If there is no further valid extension at the root of the search tree, the search terminates.

(4) *Display:* If the search path reaches an endpoint of the search tree, the algorithm displays the contents of **VCR**, which corresponds to a complete coloring of the graph.

The following procedure Color_Backtrack generates all the valid colorings of a graph $G(V, E)$ that use at most M colors. We assume G is represented as an adjacency matrix. We use a procedure Next(VCR,k), whose input is a partial coloring VCR(1..k − 1) and k. Next returns in VCR(k) the next valid coloring for vertex k which is consistent with the given partial coloring, and sets VCR(k) to $M + 1$ if there is no valid extension of the partial coloring to k. Figure 7-8 gives an example for the graph of Figure 7-5 with $M = 3$. The trace shows the first 18 iterations of the search loop.

Stage	v_1	v_2	v_3	v_4	v_5	ACTION	k
0	0	0	0	0	0	ADVANCE	1
1	1	0	0	0	0	ADVANCE	2
2	1	1	0	0	0	ADVANCE	3
3	1	1	2	0	0	ADVANCE	4
4	1	1	2	2	0	ADVANCE	5
5	1	1	2	2	3	DISPLAY	5
6	1	1	2	2	4(0)	BACKUP	4
7	1	1	2	3	0	ADVANCE	5
8	1	1	2	3	4(0)	BACKUP	4
9	1	1	2	4(0)	0	BACKUP	3
10	1	1	3	0	0	ADVANCE	4
11	1	1	3	2	0	ADVANCE	5
12	1	1	3	2	4(0)	BACKUP	4
13	1	1	3	3	0	ADVANCE	5
14	1	1	3	3	2	DISPLAY	5
15	1	1	3	3	4(0)	BACKUP	4
16	1	1	3	4(0)	0	BACKUP	3
17	1	1	4(0)	0	0	BACKUP	2
18	1	2	0	0	0	ADVANCE	3

v_i column: gives value of VCR(i) at end of stage.

k column: gives value of parameter k at end of stage.

4(0): indicates Next returns 4, which backtrack control then resets to 0, as required for BACKUP action.

Figure 7-8. Partial trace of backtrack vector for graph of Figure 7-5 where $M = 3$.

```
Procedure Color_Backtrack (G(V,E), M)

(* Lists all colorings of G using ≤ M colors *)

var  G: Graph
     M: Integer
     k:  1..|V|
     VCR(|V|): 0..M + 1
     Exit: Boolean

Set  VCR (1..|V|) to 0
Set  k to 1
Set  Exit to False

repeat

    Next(VCR,k)
    case
         (* CONDITION *)                    (* ACTION *)

    1. VCR(k) ≤ M and   k < |V|:  Set k to k + 1      (*ADVANCE*)

    2. VCR(k) ≤ M and   k = |V|:  Display VCR(1..|V|)  (*DISPLAY*)

    3. VCR(k) > M and   k > 1:    Set VCR(k)  to 0    (*BACKUP*)
                                  Set k to k- 1

    4. VCR(k) > M and   k = 1:    Set Exit  to True   (* EXIT *)

until Exit
End_Procedure_Color_Backtrack
```

7-4 FIVE-COLOR ALGORITHM

Planar graphs are four-colorable, but the proof of the Four-Color Theorem is profoundly difficult. (The current proof has over 1,000 special cases which required over 1,000 hours of computer time to analyze.) The proof of the Five-Color Theorem is rather simple and serves as the basis of an $O(|V|^2)$ five-coloring algorithm. A more elaborate $O(|V|)$ five-color algorithm for planar graphs is described in Chiba, Nishizeki, and Saito (1981).

Proof of Five-Color Theorem. The proof is by induction on $|V(G)|$ and relies on the existence of a vertex v in G of degree at most 5. We will assume that $G - v$ is five-colorable and will show how to modify and extend its coloring to a five-coloring of G. We denote the colors by $1 . . 5$.

If the neighbors of v in G are colored using only the four colors in $G - v$, we extend the coloring from $G - v$ to G by coloring v with a fifth color not used by any of its neighbors.

If the neighbors of v use all five colors, we first modify the coloring of $G - v$ so the neighbors use only four colors after which we can again extend it to a five-coloring of G. Let us define $H(i,j)$, $i,j = 1, \ldots, 5$, as the subgraph induced in $G - v$ by the vertices of $G - v$ colored i and j. For simplicity, assume the neighbors of v are arranged in clockwise order around v in some planar representation of G are $v_1, \ldots, v_5$, and that v_i has color i. There are two cases according to whether or not v_1 and v_3 lie in different components of $H(1,3)$.

v_1 and v_3 in different components of $H(1,3)$

Let v_i $(i = 1,3)$ lie in component $H_i(1,3)$ of $H(1,3)$. If we invert the colors in $H_3(1,3)$, changing every color-1 to color-3, and color-3 to color-1 and leave the colors in $H_1(1,3)$ unchanged, we obtain a valid coloring in which v_1 and v_3 are both color-1; so we are free to color v with 3.

v_1 and v_3 in same component of $H(1,3)$

There must be a path from v_1 to v_3 in $H(1,3)$. Together with the edges (v,v_1) and (v,v_3), this path determines a cycle C in G. Because of the clockwise arrangement of the neighbors of v, one of the vertices v_2 and v_4 must lie in the interior of this cycle and one must lie in its exterior. Consequently, v_2 and v_4 cannot lie in the same component of $H(2,4)$, since any path in $H(2,4)$ from v_2 to v_4 would have to pass through $C - v$, which is impossible since the vertices of $C - v$ have colors 1 or 3. Therefore, v_2 and v_4 fall under the previous case. This completes the extension of the coloring of G-v to a coloring of G, and so completes the proof of the Five-Color Theorem.

Five-color algorithm. The inductive proof leads naturally to a recursive algorithm. We can represent the graph as a linear list with vertices:

```
type  Vertex = record
                  Index: 1..|V|
                  Color: 0..5
                  Degree: 0..|V| − 1
                  Successor: Vertex pointer
                  Edge List: Edge pointer
               end
```

In addition to its usual fields, each vertex has a color, which is initially 0, and a degree. The other types are defined as usual. Next(G, w) returns in w a pointer to the record for the next vertex in G, and fails if there is none. A high level description of the procedure follows.

```
Procedure Five_Color (G)

(* Returns five-coloring of planar G *)

var   G: Header
      i, j, v, w: Vertex pointer
      C₁, C₂: Component identifier
      Nextcount: Integer function
      Next: Boolean function

if    |V(G)| ≤ 5
```

then   **while** Next(G,w) **do Set** Color(w) to Nextcount

else   Select v in V(G) of degree at most 5

       Five-Color(G − v)

       if   Adj(v) uses 5 colors

       then (* Modify coloring of G − v *)

           Find i and j in Adj(v) from distinct components C_1 and C_2 of the subgraph induced by the vertices colored the same as i and j

           for each w in C_1 **do if**   Color(w)   = Color(i)
                                then   **Set** Color(w) to Color(j)
                                else   **Set** Color(w) to Color(i)

       Set Color(v) to color not used in Adj(v)

    End_Procedure_Five_Color

     An example is shown in Figures 7-9 and 7-10.
     We used a planar representation of the graph in the proof when we assumed we knew a clockwise ordering of the vertices neighboring v, but we do not need a planar representation in the algorithm. We only need to find a pair of vertices neighboring v which are in different components of their color induced subgraph. That is, we test each pair of neighbors i and j to determine whether or not they are connected in the induced subgraph on Color(i) and Color(j). We have to test at most ten pairs since v has only five neighbors. We can perform each test in $O(|E|)$ time using depth first search. Since the graph is planar, $|E| = O(|V|)$, and so each test is actually $O(|V|)$; so all ten tests take only $O(|V|)$ time.

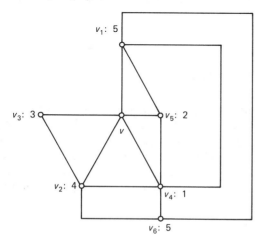

Figure 7-9. $G − v$ with recursively defined five coloring.

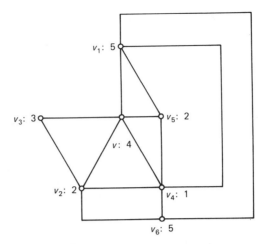

Figure 7-10. Coloring of G after modification of $G - v$.

If we denote the time required by Five_Color(G) by $f(|V(G)|)$, then f satisfies the recurrence

$$f(|V(G)|) = f(|V(G)| - 1) + c|V|.$$

The $c|V|$ term comes from the time it takes to locate and delete a vertex v of degree at most 5, and the time to find a pair of vertices i and j disconnected in the induced subgraph. Iteration gives $O(|V|^2)$ performance.

Any recursive algorithm can be unravelled into an iterative algorithm, which may be more efficient but more cluttered than the recursive version. To convert the five-color algorithm to an iterative form, we first explicitly identify the vertices $R(k)$ of degree at most five used at each stage of the recursive algorithm. We then color a small initial graph directly and extend the coloring by successively adding back the previously removed vertices $R(k)$, recoloring the intermediate graphs as we proceed, just as in the recursive algorithm.

REFERENCES AND FURTHER READING

See Behzad, et al. (1979) for a detailed discussion of coloring problems and an overview of the proof of the Four-Color Theorem. Parenthetically, the proofs of theorems in Behzad, et al. are exceptionally clear, and several have served as the basis for our own proofs. See also Appel and Haken (1977) for a popular review of their ground-breaking proof. A faster (linear) time version of the five-color algorithm is given in Chiba, Nishizeki, and Saito (1981). The approximate coloring algorithm described in the exercises, and a generalization, is given in Wigderson (1983). See Capobianco and Molluzzo (1978) for interesting examples, such as on the unique coloring problem.

EXERCISES

1. Determine the vertex and edge chromatic numbers of each of the five regular polyhedra.
2. Let $G(V, E)$ be a connected cubic graph of order >4, and girth $= 3$. Determine the VCHR(G).
3. Let G be bipartite. Prove that the vertex chromatic number of the complement of G equals the order of the maximum order complete subgraph in G^c
4. A graph $G(V, E)$ is said to be *uniquely n-colorable* if VCHR(G) equals n, and every n-coloring of G induces the same partition of the vertices of G. Prove that if G is uniquely n-colorable, G is $(n - 1)$-connected.
5. Prove that if G is bipartite, ECHR(G) $=$ max(G).
6. A graph $G(V, E)$ embedded in the plane is said to be *n-region-colorable* if its regions can be colored with at most n colors, so that adjacent regions are colored differently. Prove that every planar graph is four-colorable if and only if every planar embedding of a graph is *four-region-colorable*.
7. Prove that if G is connected, not an odd cycle, and all its cycles have the same parity, ECHR(G) $=$ max(G).
8. Prove that if G is a nontrivial regular graph of odd order, ECHR(G) $=$ max(G) $+ 1$.
9. Write a program that simulates the operation of a zero-knowledge password system using the graph passwords method described in the text.
10. Write a program that performs constrained scheduling and uses a backtracking algorithm to find an optimal schedule.
11. Use Monte Carlo estimation (see Chapter 2) to estimate the performance of a backtracking algorithm for finding the k-colorings of a graph. Compare the number of vertices in the estimated backtrack tree for the colorings with the number of vertices in the actual backtrack tree for the colorings. Examine the dependence of the performance on the order in which the vertices are scanned. Can you suggest a rationale for a preferred order for scanning?
12. Suppose each of the edges of $K(6)$ is colored either red or blue. Prove the resulting graph contains a monochromatic triangle, that is, a cycle of length 3 all of whose edges have the same color.
13. Implement the recursive version of the five-color algorithm.
14. The following algorithm approximately colors a three-chromatic graph $G(V, E)$.

    (a) Set n to $|V|$ and color to 1.
    (b) while max(G) $\geq n^{(1/2)}$
           do   Select a vertex v of maximum degree
                Color the Adj(v) with color and color $+ 1$
                Increment color by 2
                Set color(v) to color
                Set G to G $- $ v $- $ Adj(v).

(c) Use the greedy procedure given for the weakened form of Brooks' Theorem to color any remaining vertices.

Give the performance of this algorithm and prove it colors a three-chromatic graph with at most $3n^{(1/2)}$ colors. What happens if the algorithm is applied to a k-chromatic graph? (See Wigderson (1983) for a more general discussion.)

8

Covers, Domination, Independent Sets, Matchings, and Factors

8-1 BASIC CONCEPTS

We now introduce the elements of the theory of graph covering and related invariants. If $G(V, E)$ is a graph, then a *vertex* of G is said to *cover any edge of G it is incident with,* and similarly, an *edge* of G is said to *cover its endpoints*. A set of vertices in G that covers every edge of G is called a *vertex cover,* while a set of edges in G that covers every vertex of G is called an *edge cover*.

Domination is a subtly different notion. While a *vertex* is said to cover an incident edge, it is said to *dominate* both itself and any adjacent vertex. A *vertex dominating set* is a set of vertices S such that every vertex in G is dominated by some vertex in S. Similarly, an *edge* is said to *dominate* the edges incident to its endpoints, as well as itself. An *edge dominating set* is a set of edges X such that every edge in G is dominated by some edge in X.

The notion of an *independent set* is complementary to that of a cover. While a cover requires adjacency, an independent set requires nonadjacency. We will call a set of vertices no two of which are adjacent an *independent set of vertices*. We will call a set of edges no two of which are adjacent an *independent set of edges* or a *matching*. An independent set of vertices is said to be *maximal* if any vertex not in the set is dominated by a vertex in the set.

There are simple relations between the extreme values of the sizes of covers, dominating sets, and independent sets. We use the following notation.

Minimum cardinality of a vertex cover of G: Min-VCov(G)
Minimum cardinality of an edge cover of G: Min-ECov(G)
Minimum cardinality of a vertex dominating set: Min-VDom(G)
Minimum cardinality of an edge dominating set: Min-EDom(G)
Maximum cardinality of a matching on G: Max-Match(G)
Maximum cardinality of a vertex independent set of G: Max-Indep(G)

Each of these invariants has a simple physical interpretation. For example, suppose we can install "service centers" at a subset of the vertices of a graph, where each center can provide service both for its home vertex and adjacent vertices. Then, the minimum number of such centers needed to provide service to the whole graph equals Min-VDom(G). The following theorems are easily proved.

Theorem (Covers, Dominating Sets, Matchings, and Independent Sets). If $G(V, E)$ is a graph with no isolated vertices, then

(1) Min-EDom(G) $\leq$ Max-Match(G) $\leq$ Min-VCov(G), and

(2) Min-VDom(G) $\leq$ Max-Indep(G) $\leq$ Min-ECov(G).

As an example, consider the proof of Min-VDom(G) $\leq$ Max-Indep(G). Let S be an independent set of vertices of G such that $|S|$ = Max-Indep(G). If there exists a vertex v in V-S which is not adjacent to some vertex of S, then $S \cup \{v\}$ is an independent set of cardinality Max-Indep(G) + 1. Therefore, S must be a vertex dominating set, whence it follows that Min-VDom(G) $\leq$ Max-Indep(G).

It follows from this theorem that if the size of a matching equals the size of a vertex cover, the matching is a maximum matching and the vertex cover is a minimum vertex cover. Similar remarks hold for the other quantities.

Theorem (Gallai's Complementarity Relations). If $G(V, E)$ is a graph containing no isolated vertices, then

(1) Max-Match(G) + Min-ECov(G) = $|V(G)|$, and

(2) Max-Indep(G) + Min-VCov(G) = $|V(G)|$.

To prove (1), we will first show that Max-Match(G) + Min-ECov(G) $\leq$ $|V(G)|$. Let E_{ind} be an independent set of edges of G of cardinality Max-Match(G). Thus, E_{ind} covers 2 Max-Match(G) vertices of G. For each vertex in G not covered by an edge of E_{ind}, select an additional edge from $E(G)$ and denote the union of these edges by E_{extra}. Then, $E_{ind} \cup E_{extra}$ is an edge cover of G, so that $|E_{ind} \cup E_{extra}|$ is at least Min-ECov(G). But, $|E_{ind}| + |E_{ind} \cup E_{extra}| = |V|$; so Max-Match($G$) + Min-ECov($G$) $\leq$ $|V(G)|$. To prove that Max-Match(G) + Min-ECov(G) $\geq$ $|V(G)|$, let E_{min} be a minimum edge cover of G. Then, the subgraph determined by E_{min} is acyclic. If we select an edge from each of its components, and denote the resulting set of edges by E_{comp}, then $|E_{comp}| \leq$ Max-Match(G), and $|E_{min}| + |E_{comp}| = |V|$. It follows that Max-Match(G) + Min-ECov(G) $\geq$ $|V(G)|$, which completes the proof of (1).

To prove (2), first observe that S is an independent set of vertices of $G(V, E)$ if and only if V-S is a vertex cover of G. Then, let S_{max} be a maximum independent set, and let T_{min} be a minimum vertex cover of G. By observation, $S' = V - S_{max}$ is a vertex cover, and $T' = T - T_{min}$ is an independent set. Therefore,

$$|S'| = |V - S_{max}| = |V| - \text{Max-Indep}(G) \geq \text{Min-VCov}(G),$$

while,

$$|T'| = |V - T_{min}| = |V| - \text{Min-VCov}(G) \leq \text{Max-Indep}(G).$$

Combining the inequalities, we obtain (2).

Factors.   A *factor* of a graph $G(V, E)$ is a spanning subgraph of G. If G can be expressed as the edge sum of factors of G, the edge sum is called a *factorization* of G. An *r-factor* of a graph $G(V, E)$ is an r-regular factor of G. Thus, a 1-factor of G is a perfect matching of G, while a 2-factor of G is either a spanning cycle of G, or a disjoint union of cycles which together span G. G is said to be *r-factorable* if it has a factorization consisting only of r-factors. Koenig's Theorem (Section 1-4) shows regular bigraphs are 1-factorable. The following theorems are famous.

Theorem (Tutte's 1-factor Condition).   A nontrivial graph $G(V, E)$ has a 1-factor if and only if for every proper subset S of $V(G)$, the number of components of G-S of odd order is at most $|S|$.

Theorem (Petersen's 2-factor Condition).   A nonempty graph $G(V, E)$ is 2-factorable if and only if G is $2r$-regular for some $r \geq 1$.

Theorem (Petersen's Characterization of Cubic Graphs).   Every bridgeless cubic graph can be expressed as the edge sum of a 1-factor and a 2-factor.

The Petersen graph (see Section 1-4) shows not every bridgeless cubic graph is 1-factorable.

There are noteworthy special results on covers and matchings for bigraphs. We have already established (Section 1-3) a noncontracting condition for the existence of a spanning matching on a bigraph. The following theorem strengthens inequalities we have given for general graphs.

Theorem (Bipartite Matching, Covering, and Independence).   If $G(V, E)$ is a bipartite graph with no isolated vertices, then

(1) Max-Match(G) = Min-VCov(G), and

(2) Max-Indep(G) = Min-ECov(G).

We shall only prove statement (1). By the inequalities for the general theorem, we already know that Max-Match(G) $\leq$ Min-VCov(G); so we only need to prove that Max-Match(G) $\geq$ Min-VCov(G). Denote the bipartite parts of G by V_1 and V_2, and let M be a maximum matching of cardinality Max-Match(G). If G has a perfect matching, then Max-Match(G) = Min-VCov(G), and we are done.

Otherwise, let X be the set of vertices in V_1 not covered by M. Thus, $|M| = |V_1| - |X|$. Let R be the set of vertices in G reachable from X by alternating paths with respect to M. Let W_i ($i = 1, 2$) be the part of R in V_i. By the definition of X, $W_1 - X$ is matched by M to W_2, so $|W_2| = |W_1| - |X|$, and W_2 is contained in **ADJ**(W_1). Conversely, W_1 is contained in ADJ(W_2), since G contains an alternating path from X to w, for every vertex w in ADJ(W_1). Therefore, ADJ(W_1) = W_2, and $|$ADJ$(W_1)| = |W_2| = |W_1| - |X|$. Therefore, the set $(V_1 - W_1) \cup W_2$ is a vertex cover of cardinality $|M|$, since otherwise there is an edge (w_1, y) with w_1 in W_1 and y not in W_2, contrary to the conclusion that ADJ(W_1) = W_2. This completes the proof of the theorem.

Regular bigraphs are 1-factorable by Koenig's theorem, although arbitrary bigraphs may not be. However, bigraphs can always be partitioned into a small number of edge disjoint matchings.

Theorem (Matching Partition of Bipartite Graphs). The edges of a bipartite graph $G(V, E)$ can be partitioned into $\max(G)$ disjoint edge matchings.

The matchings guaranteed by this theorem are not necessarily spanning matchings, so this is not a factorization result. It follows from the theorem that if G is a bigraph, then $\text{ECHR}(G) \leq \max(G)$.

8-2 MODELS

8-2-1 Independence Number and Parallel Maximum

We have previously given a theoretical lower bound on the speed with which n numbers could be sorted using a sequential binary comparison sort (Chapter 4). We will now establish a lower bound on the speed with which the largest of n numbers can be found using k parallel processors. The model for the problem is an undirected graph in which the result of a round of comparisons is represented by an independent set of vertices. An optimal maximum value selection algorithm based on the lower bound analysis is also presented.

The model of parallel computation is CREW PRAM. That is, we assume the k processors can compare k not necessarily distinct pairs of values in constant time (see Chapter 9 for further discussion of shared memory models of parallel computation). The processors may read common memory locations concurrently, but make no concurrent writes to common locations. For simplicity, we assume the n numbers are distinct and the number k of processors equals n.

Let A denote an arbitrary binary comparison algorithm that uses k processors to find the largest value in a set M of n numbers. The algorithm consists of a sequence of constant time steps, where each of the k processors makes a comparison, in parallel. Let M_i denote the subset of values in M that have not yet been shown to be smaller than any of the other values in M by the end of step i. In other words, each value in M_i remains a candidate for the maximum value at the next step $i + 1$ of the algorithm. However, every value in $M - M_i$ has definitely been excluded from being the maximum by step $i + 1$. Naturally, the algorithm terminates when $|M_i| = 1$.

To prove the lower bound on the termination time, we require the following extremal result.

Theorem (Minimax Independent Set of Turan). Let $G(V, E)$ be a graph of order p and size k, and let r be the order of the largest vertex independent set in G. Then,

(1) r is at least $p^2/(2k + p)$. Furthermore, if we define the graphs $H(p, r)$, for positive integers p and r ($r < p$), as the disjoint union of $r - rt + p$ complete graphs $K(t)$ and $rt - p$ complete graphs $K(t - 1)$, where t is the smallest integer as large as p/r, then,

(2) $H(p, r)$ has size $(t - 1)(2p - rt)/2$, and has the minimum number of edges among all graphs of order p with vertex independence number at most r.

See Turan (1954) for a proof. The desired theoretical lower bound on the speed of parallel maximum selection is then given by the following theorem.

Theorem (Lower Bound for Parallel Maximum). Any algorithm based on binary comparisons and using n processors to find the maximum of n elements takes at least time $O(\log(\log(n)))$. The bound is sharp.

The proof is as follows. Consider the state of affairs at the beginning of step $i + 1$ of the algorithm X. At this point, A can compare values from M_i either to each other or to values from $M - M_i$, or compare values from $M - M_i$ to each other. Comparisons between $M - M_i$ values can yield no new information about the maximum, because every one of these values is already known by the end of step i to be excluded as a candidate for maximum. On the other hand, we cannot a priori exclude the possibility that in each comparison between a value from $M - M_i$ and a value from M_i, the value from M_i is greater; so these comparisons might also yield no additional information that could narrow the list of candidates for maximum. We will establish a lower bound under the assumption that this worst case occurs. Consequently, we only need to consider the effect that comparisons among values in M_i can have on the number of candidates for maximum.

Observe that at each step, the algorithm makes at most k (or n) comparisons. We model these comparisons by a graph $G(V, E)$ where

(1) $V(G)$ has $|M_i|$ vertices, one for each of the current candidates for maxima M_i, and
(2) $E(G)$ has at most k edges, one for each of the binary comparisons made at step i.

The size of M_{i+1}, the updated list of candidates for maximum, is related to a feature of the model graph. Thus, suppose S is an independent set of vertices in G. S corresponds to a set of numbers no two of which are compared by the algorithm at step i. If it happens that each of the elements in S is greater than every value in $M_i - S$, any of the numbers (vertices) in S could be the maximum. It follows that $|M_{i+1}|$ could be (though not necessarily must be) as large as the largest set of independent vertices in any graph of order $|M_i|$ and size k (or n). Therefore, $|M_{i+1}|$ must be at least as large as the smallest such independent set taken over all such graphs. That is,

$$|M_{i+1}| \geq \min \{\text{max cardinality of an independent set in } G'\},$$

where the minimum is taken over all graphs G' of order $|M_i|$ and size at most n.

Applied to our problem, Turan's theorem (1) implies that

$$|M_{i+1}| \geq |M_i|^2/(2k + |M_i|).$$

The lower bound for the value of i on termination follows readily. $|M_0|$ must equal n, since M_0 is just the original set of n values. The algorithm cannot stop until $|M_i| = 1$. Therefore, the earliest time i at which the algorithm can terminate satisfies

$$i \geq \log(\log(n)) - c,$$

for some constant c, which completes the proof of the lower bound of the theorem. It remains to show the bound is sharp.

The preceding part of the proof demonstrates that the parallel maximum cannot be computed in faster than $O(\log(\log(n)))$ time using n processors. A related result is obtainable if $k < n$ processors are used. Using part (2) of Turan's Theorem, we can show the lower bound is attainable. For simplicity, we again assume $k = n$. We will design an optimal parallel algorithm for finding the maximum using a special class of extremal graphs to determine which elements to compare at each step. The extremal comparison graphs will be such that $|M_i|$ is reduced to 1 in $\log(\log(n))$ steps.

The extremal graphs are just the graphs $H(p, r)$ of Turan's Theorem. $H(p, r)$ consists of r disjoint complete subgraphs. The comparisons for each complete subgraph yield exactly one potential maximum, or r candidates for maximum altogether. We will select a feasible $H(p, r)$ with the correct number of vertices and edges that minimizes r. The feasible graphs are $H(|M_i|, r)$ of order $|M_i|$ and size at most k. By Turan's Theorem, or directly by the definition of these graphs, the feasible graphs are those for which $(t - 1)(2|M_i| - rt)/2 \le k$, where t is the smallest integer as large as $|M_i|/r$. We select that $H(p, r)$ of order p and size at most k that minimizes r. If we use this choice of extremal graph to determine the comparisons at step i, the number of potential maxima at step $i + 1$ is bounded above by

$$|M_{i+1}| \le \min \{r \,|\, (t - 1)(2|M_i| - rt)/2 \le k\} .$$

which implies,

$$|M_{i+1}| \le |M_i|^2/(ck) ,$$

for some constant c. Since $|M_0|$ equals n, $|M_i|$ is reduced to 1 for some i less than $\log(\log(n)) + c$. This proves the sharpness of the lower bound.

8-2-2 Matching and Processor Scheduling

There is a simple relationship between maximum matchings and optimal schedules for a system of tasks executing on a pair of parallel processors. Thus, suppose that each of a set of tasks takes unit time to execute and that the order in which the tasks can be scheduled is restricted by precedence constraints. The precedence constraints are represented by an acyclic digraph $G(V, E)$, whose vertices correspond to the set of tasks, and whose edges represent the precedence constraints. We will show how to schedule these tasks in parallel on a pair of processors in such a way as to minimize the earliest completion time of the system. Then, we shall bound the completion time in terms of the size of a maximum matching on G.

The optimal scheduling algorithm is as follows. We first rank the vertices of $G(V, E)$ according to a priority scheme defined as follows.

(1) Identify tasks T_{i1} to T_{ik} having no succcessors and assign them priorities 1 to k, respectively. Set Next to $k + 1$.

(2) Repeat (2a and b) until every task has been assigned a priority.

    (2a) For each task T all of whose successor tasks $s_1, \ldots, s_m$ have been assigned a priority, let $v(T)$ be an m-vector with components as the priority labels of the tasks in ascending order.

    (2b) Select the lexicographically smallest vector-labelled task T, and set priority(T) to Next. Then, set Next to Next + 1.

Figure 8-1 illustrates the procedure.

Once the priorities have been assigned, we schedule the processors according to the following rule.

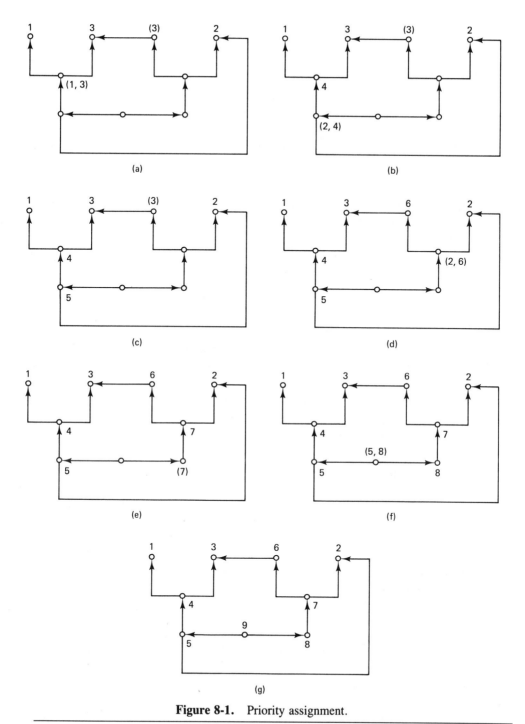

Figure 8-1. Priority assignment.

Whenever a processor becomes free, assign it to execute the highest priority task remaining to be completed.

Figure 8-2 gives an optimal schedule for the example of Figure 8-1.

The proof of the correctness of this scheduling algorithm is fairly difficult. We refer to Coffman and Denning (1973). However, it is easy to prove the following theorem.

Theorem (Scheduling Lower Bound).   Let a system of tasks constrained by a set of precedence constraints represented by a digraph $G(V, E)$ be defined as shown. Let t_{min} denote the minimum time in which the task system can be completed. Then,

$$t_{min} \geq |V(G)| - |M(C(U(T(G))))|.$$

where $M()$ returns a maximum matching on its argument graph, $C()$ returns the complement of its argument, $U()$ returns the undirected graph underlying its digraph argument, and $T()$ returns the transitive closure of G.

The proof of the bound is simple. Observe that there is an edge between any pair of tasks in $U(T(G))$ which are either directly or indirectly dependent on one another. Therefore, $C(U(T(G)))$ contains edges only between those tasks that are completely independent of one another. Consequently, if an optimal processing schedule schedules the pairs of tasks $T1_i, T2_i, i = 1, \ldots, k$ in parallel, the set of edges $\{(T1_i, T2_i)\}$ determines a matching in $C(U(T(G)))$. Therefore, the maximum possible degree of parallelism attainable is constrained by the size of a maximum matching in $C(U(T(G)))$. If a maximum matching has size m, at most m tasks can be parallelized. It follows that the optimal schedule must take time at least $|V(G)| - m$. This completes the proof.

Refer to Figure 8-3 for an example. We observe without proof that the optimal scheduling algorithm attains the lower bound. Observe also that not every maximum matching determines a feasible schedule. For example, consider the matching $\{(T_1, T_4), (T_2, T_3)\}$ for the graph in Figure 8-3b.

8-2-3 Degree Constrained Matching and Deadlock Freedom

A deadlock avoidance algorithm allocates a set of shared resources among a set of processes in a way that avoids even the possibility of a deadlock. In contrast, a system of processes and resources is said to be *deadlock free* if no deadlock is possible, regardless of how the resources are allocated. Since no special measures are needed to prevent deadlock, the overhead of a deadlock avoidance algorithm is avoided. We will describe a generalized matching property for a resource allocation graph that guarantees a system is deadlock free. But, first we will introduce a bigraph model of the resource requirements of a system of processors and processes.

Consider a system with a set P of n processes $P_i, i = 1, \ldots, n$, and a set R of m types of resources $R_j, j = 1, \ldots, m$, where we assume the resources are serially reusable, $Avail(j)$ units of resource j are available for allocation, and $Claim(i, j)$ gives the

Time	Available	P_1	P_2
0	{9}	9	—
1	{5, 8}	5	8
2	{4, 7}	4	7
3	{1, 2, 6}	6	2
4	{1, 3}	1	3

Figure 8-2. Optimal schedule for Figure 8-1.

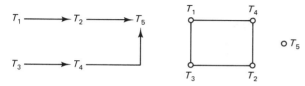

(a) Task precedence digraph G.   (b) Associated graph $C(U(T(G)))$.

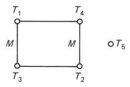

Associated optimal schedule

Time = 0: T_1, T_3

Time = 1: T_2, T_4

Time = 3: T_5

(c) Associated graph $C(U(T(G)))$.   **Figure 8-3.** Lower bound example.

maximum number of units of resource j which process i might possibly need at any point. The *shortage of resource j relative to a subset of processes S* is defined by

$$Short(S,j) = \sum_{i \in S} Claim(i,j) - Avail(j).$$

$Short(S,j)$ represents the excess of demand for resource j from the processes in S over the total supply of j available in the system. If $Short(S,j) < 0$, there can never be a shortage of resource j, so the system can never be blocked on j. We are interested in conditions that guarantee the system is never blocked even when $Short(S,j) > 0$, for some resources j.

The resource requirements of the system are modeled by a shortage graph which summarizes the process-resource claims and potential resource shortages. The *shortage graph* for a system $(P, R, Claim, Avail)$ *with respect to a subset of processes S in P* is a labelled bipartite graph denoted by $Shortage(S, P, R, Claim, Avail)$ (or $G(S)$ for *short*) with parts V_1 and V_2 such that

(1) There is a vertex in $V_1(G)$ corresponding to each process in S, and a vertex in $V_2(G)$ for each resource class in R;

(2) There is an edge (p, r) between process vertex p and resource vertex r if and only if

   (2a) $Claim(p, r) > 0$, that is, process p has a potential claim on an instance of resource r, and

   (2b) $Short(S, r) > 0$, that is, there is a shortage of resource r relative to S;

(3) Each vertex has a label. $Label(p)$ equals 1, for every process vertex p; $Label(r)$ equals the potential shortfall of r: $\max\{0, Short(S, r)\}$, for every resource vertex r.

Refer to Figure 8-4 for an example.

Let us define a *degree constrained matching M* for a labelled graph $G(V, E)$ as a set M of edges of G such that the number of edges of M incident at any vertex v of G

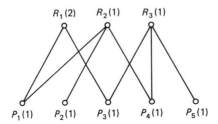

	R_1	R_2	R_3
P_1	2	1	0
P_2	0	2	0
P_3	2	0	1
P_4	0	2	1
P_5	0	0	2

Avail(1) = 2
Avail(2) = 4
Avail(3) = 3

(a) **Claim** matrix and **Avail**.

$R_1(2)$ $R_2(1)$ $R_3(1)$

$P_1(1)$ $P_2(1)$ $P_3(1)$ $P_4(1)$ $P_5(1)$

(b) Shortage graph **Shortage** (*P, P, R*, **Claim, Avail**).   **Figure 8-4.**   Example shortage graph.

is $\leq Label(v)$. Refer to Figures 8-5a and 8-6a for examples. A *maximum degree constrained matching* is a degree constrained matching of maximum cardinality. If there exists a degree constrained matching for the shortage graph $G(S)$ that matches every process vertex in $G(S)$, then $G(S)$ is said to have a *complete degree constrained matching*.

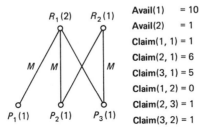

$R_1(2)$ $R_2(1)$

M M M

$P_1(1)$ $P_2(1)$ $P_3(1)$

Avail(1)   = 10
Avail(2)   = 1
Claim(1, 1) = 1
Claim(2, 1) = 6
Claim(3, 1) = 5
Claim(1, 2) = 0
Claim(2, 3) = 1
Claim(3, 2) = 1

(a) Shortage graph with complete matching.

Alloc(1, 1)   = 0   Alloc(1, 2)   = 0
Alloc(2, 1)   = 5   Alloc(2, 2)   = 1
Alloc(3, 1)   = 5   Alloc(3, 2)   = 0
Request(1, 1) = 1   Request(1, 2) = 0
Request(2, 1) = 1   Request(2, 2) = 0
Request(3, 1) = 0   Request(3, 2) = 1

Notation:
Claim(*i, j*):   amount of resource *j* claimed by process *i*
Alloc(*i, j*):   amount of resource *j* allocated to process *i*
Request(*i, j*): amount of resource *j* requested by process *i*

(b) Deadlock state.

Figure 8-5. Constructing deadlock configuration (b) from a complete matching in (a).

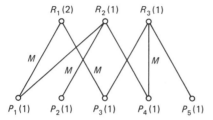

(a) Maximum degree constrained matching for $G(P)$ of Figure 8-4.

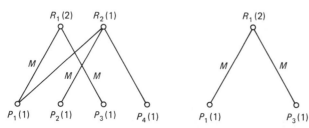

(b) First reduced graph.

(c) Second reduced graph.

Figure 8-6. Testing for deadlock freedom.

We can characterize deadlock free systems in terms of complete degree constrained matchings.

Theorem (Characterization of Deadlock Freedom). Let $P = \{P_i, i = 1, \ldots, n\}$ be a set of processes, and let $R = \{R_j, j = 1, \ldots, m\}$ be a set of resource types, with *Claim* and *Avail* as defined above. Then, the system is deadlock free if and only if there is no subset S of P such that $Shortage(S, P, R, Claim, Avail)$ has a complete degree constrained matching.

We prove just the only if part of the theorem. Suppose a set of processes S exists for which $Shortage(S, P, R, Claim, Avail)$ has a complete degree constrained matching M. Then, we can construct a deadlock state of the system as follows. Suppose there are m_1 edges of M incident at a particular resource vertex r. By definition, each process incident on a matching edge of r has a nonzero potential claim against r. Let us assign a request of one unit of r to each such process, for a total of m_1 requested units. Since the number of matching edges incident at r is at most equal to $Short(S, r)$ by the degree constrained nature of the matching, it follows by the definition of $Short(S, r)$ that

$$Avail(j) \leq \sum_{i \in S} Claim(i, j) - m_1.$$

It follows from this that there is sufficient outstanding demand for r from S to deplete all the available supply of r. Therefore, let us assume that all $Avail(j)$ units of r have already been distributed among all those processes in S that have nonzero claims against r, whether matched with r or not, and in such a way that none of the processes matched with r is allocated its full possible claim, thus allowing for a unit request for each of the matched processes. The resulting configuration blocks every process matched with r, since each such process is waiting for a unit of r, and all units of r have already been granted. If we construct a similar pattern of allocated claims and pending requests for every process in S; then in the resulting configuration, each process is blocked waiting for a resource held by another blocked process, that is, the system is deadlocked. This completes the proof of necessity. We refer to Kameda (1980) for the complete discussion. Figure 8-5 gives an example.

In order to define an algorithm that tests if a system is deadlock free, we need the following theorem.

Theorem (Deadlock-able Processes). Let $P = \{P_i, i = 1, \ldots, n\}$ be a set of processes, and let $R = \{R_j, j = 1, \ldots, m\}$ be a set of resource types, with *Claim* and *Avail* as defined above. Then, if a process P_i can become deadlocked, then P_i must be matched to some resource vertex in every maximum degree constrained matching of $G(P)$.

We refer to Kameda for the proof. An algorithm that tests if a system is deadlock free follows. Step (1) of the algorithm computes the shortage variables and deletes unrestricted resources. Step (3) is justified by the characterization theorem, while the iterated steps in (4) are each justified by the Theorem on Deadlockable Processes.

(1) Compute $Short(P, j)$ for every resource R_j, and discard those for which $Short(P, j)$ is negative. Then, construct the model bigraph $G(P)$ on the remaining resources.

(2) Find a maximum degree constrained matching on $G(P)$.

(3) If the matching is complete, then the system is not deadlock free.

(4) Otherwise, repeat the following step as many times as possible:
Select any process vertex p not in the matching. Delete p and every resource vertex incident with p.

(5) The system is deadlock free, if, upon completion of the step (4) iterations, no process vertex remains.

An example is shown in Figure 8-6. Figure 8-6a gives an incomplete maximum matching (p_5 is not matched). The initial iteration of step (4) deletes P_5 and R_3 (Figure 8-6b). This leaves P_4 unmatched; so p_4 is deleted in the next iteration of (4) (Figure 8-6c). At this point, every process vertex is matched; so no further iterations of (4) are possible. Since the final graph is nonempty, the system is not deadlock free.

8-3 MAXIMUM MATCHING ALGORITHM OF EDMONDS

We introduced the alternating paths algorithm for maximum matchings on bipartite graphs in Chapter 1. We now describe an alternating paths algorithm for maximum matchings on arbitrary graphs. The algorithm is complicated in the general case because the graph may contain cycles of odd length. See Figure 8-7a for an example of what can go wrong. If we start an augmenting search tree at Root, we may not detect the augmenting path Root-v_1-v_3-v_4-v_2-v_5, which gives the augmented matching shown in Figure 8-7b: it depends on the order in which we happen to scan the vertices. However, if we collapse an odd cycle as soon as it is detected, as shown in Figure 8-7c, we can recognize the augmenting path in the collapsed graph. If we then merely follow that augmenting path (from v_5 back to Root), using the even length part of the odd cycle, we get an augmenting path in G, which yields the maximum matching in Figure 8-7b.

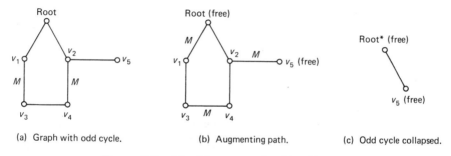

Figure 8-7. Matching on graph with odd cycle.

The same idea works for arbitrary graphs, though there are considerable variations depending on how the odd cycles (or *blossoms,* as they are called) are handled. In the most straightforward approach, where the blossoms are collapsed, and re-expanded later on demand if they lie on an augmenting path, the algorithm takes $O(|V|^4)$ time. A variation due to Gabow (1976) avoids the explicit collapse and expansion and has time $O(|V|^3)$. Just as in the binary case, the search tree can become blocked, in which case, the blocked (or *Hungarian*) search (sub)tree can be ignored in *all* subsequent searches.

Our algorithm uses the straightforward approach. A blossom B is recognized when an edge between a pair of outer vertices is scanned. It consists of that edge, plus the two tree paths from the unique common ancestor of the endpoints of the edge. The blossom is handled by shrinking it into a super-vertex b, which is labelled as free, if it contains the root of the tree, and always as an outer vertex, to ensure it is scanned from in the collapsed graph $G(b)$, obtained from G by shrinking the blossom B into a single vertex b. Thus, corresponding to any vertex v in $V(G)-V(B)$ which is adjacent to a vertex w in $V(B)$, there is a vertex v in $G(b)$ which is adjacent to b in $V(G(b))$. Any self-loops, caused by mutually adjacent vertices in $V(B)$ in G, are deleted in $G(b)$. Any parallel edges, caused if more than one vertex in B is adjacent to a common vertex outside $V(B)$, are also removed. Thus, $G(b)$ is a simple loop-free graph.

A high level view of the augmenting search tree algorithm is given in Find_ Augmenting_Search_Tree. It is similar to the bipartite search tree algorithm, except that it includes a blossom recognition and collapse step. Also, the status of the vertices (inner or outer) has to be explicitly recorded, because of the different actions that are taken if the vertex being scanned is outer or inner. The function Next(u,v) returns the next unexplored edge (u, v) at u, and fails if there is none. If Next succeeds, Find_ Augmenting_Tree either recognizes v as an augmenting vertex or determines that v is matched and so extends the search tree to v and its neighboring matching vertex, or recognizes that the edge (u, v) determines a blossom. Find_Augmenting_Tree terminates when either the search becomes blocked or when an augmenting vertex is found.

A suggestive outline of the driver algorithm is given in the procedure Maximum_Matching. The procedure Clear merely resets positional pointers and status flags, but does not have to reestablish the graph as it was before the blossoms off the augmenting path were introduced. Remove_Tree(G) just removes a Hungarian tree from G.

Get_Augmenting_Path extracts the augmenting path detected by the search algorithm. If necessary, it restores the part of G along the augmenting path from v to Root, by reexpanding any blossoms introduced along that path by Find_Augmenting_Tree.

Since blossoms can be nested and can include the original root, this process can be quite complicated. But, the basic idea is simple. Whenever a blossom is recognized, as we backtrack along the augmenting path found by Find_Augmenting_Tree, we expand it and use the even length part of the blossom between the vertex where we enter the blossom, and the vertex from which we exit the blossom, as a part of the expanded augmenting path. Refer to Figure 8-8. Observe that only blossoms lying on the augmenting path have to be expanded. A blossom off the path merely remains part of the graph as it stands for the next search phase, and has to be expanded only if it becomes an Inner vertex with respect to some future search. A detailed example is shown in Figure 8-9.

Function Find_Augmenting_Tree (G,Root,v)

(* Tries to find an augmenting path from Root to v, or fails *)

var G: Graph
       v, w, Root, Q: Vertex pointer
       Head: Vertex pointer function
       Next, Find_Augmenting_Tree, Empty: Boolean function

Set Find_Augmenting_Tree to False

Create(Q); Enqueue(Root,Q)

repeat

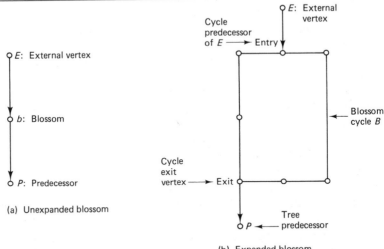

(a) Unexpanded blossom

(b) Expanded blossom

E:	Vertex external to blossom with predecessor in blossom
b:	Blossom vertex prior to expansion
P:	Search tree vertex which is predecessor of blossom b
Entry:	Vertex in expanded blossom which is predecessor of E
Exit:	Vertex in expanded blossom whose predecessor is P

Figure 8-8. Blossom structure.

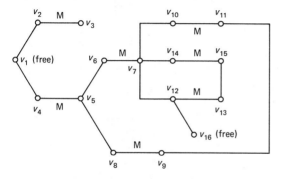

(a) Graph with initial matching and free vertices v_1 and v_{16}.

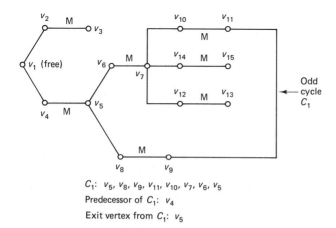

C_1: v_5, v_8, v_9, v_{11}, v_{10}, v_7, v_6, v_5

Predecessor of C_1: v_4

Exit vertex from C_1: v_5

(b) Odd cycle detected on scanning edge (v_9, v_{11}).

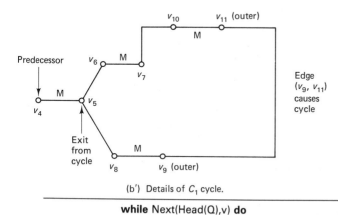

(b') Details of C_1 cycle.

Figure 8-9. Edmonds' maximum matching algorithm example.

while Next(Head(Q),v) **do**

    if v free **and** <> Root **then** (* Augmenting path found *)

        Set Pred(v) to Head(Q)

        Set Find_Augmenting_Tree to True

        Return

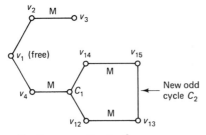

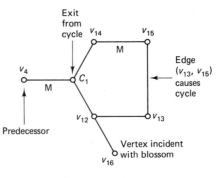

C_2: C_1, v_{12}, v_{13}, v_{15}, v_{14}, C_1

Predecessor of C_2: v_4

Exit vertex from C_2: C_1

(c) C_1 collapsed and new odd cycle C_2 detected.

(c′) Details of C_2 cycle.

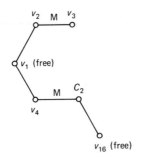

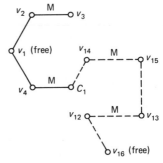

(d) Augmenting vertex v_{16} detected via edge (C_2, v_{16}) (= (v_{12}, v_{16})).

Even part of C_2 (dashed) from entry at v_{12} to exit at C_1: v_{12}, v_{13}, v_{15}, v_{14}, C_1

(e) C_2 expanded and even part of C_2 used.

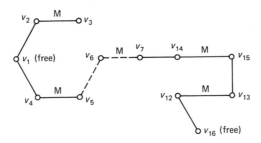

Even part of C_1 (dashed) from entry at v_7 to exit at v_5: v_7, v_6, v_5

Expanded augmenting path: v_1, v_4, v_5, v_6, v_7, v_{14}, v_{15}, v_{13}, v_{12}, v_{16}

(f) C_1 expanded with the even part of C_1 used.

(**Figure 8-9** continued)

---

else if v unscanned   **then** (* Extend search tree *)
        Let w be vertex matched with v

        Set Pred(v) to Head(Q)
        Set Status(v) to Inner

                    Set Pred(w) to v
                    Set Status(w) to Outer

                    Enqueue(w,Q)

        else if v Outer   **then** Create_Blossom (G,Head(Q),v)

        else (* v Inner *)   **then** (* Ignore *)

    until Empty(Dequeue(Q))

    End_Function_Find_Augmenting_Tree

    Procedure Maximum_Matching (G)

    (* Finds a maximum matching in G *)

    var G: Graph
            Root, v: Vertex pointer
            Next_Free, Find_Augmenting_Tree: Boolean function

    while   Next_Free (G,Root) **do**

      if   Find_Augmenting_Tree (G, Root, v)

      then   Get_Augmenting_Path (G,Root,v)
               Apply_Augmenting_Path(G,Root,v)
               Clear(G)

      else   Remove_Tree (G)

    End_Procedure_Maximum_Matching

The proof of the correctness of the exclusion of Hungarian trees from subsequent searches is the same as in Chapter 1. The correctness of the blossom management procedures is established by the following theorem.

Theorem (Blossom Management).   The Find_Augmenting_Tree algorithm succeeds if and only if G has an augmenting path.

The sufficiency is clear from the blossom expansion techniques that we have described. Therefore, it only remains to prove that if there is an augmenting path in the original graph, then some such path must be found by Find_Augmenting_Tree. Let us first observe that if the search tree ever becomes blocked, any adjacent outer vertices are, by that point, collapsed into a common blossom, as follows directly from the nature of the search procedure. Suppose now that Find_Augmenting_Tree fails, even though there is an augmenting path $v_0, \ldots, v_m$ (where m is necessarily odd) in the original graph G. It will be helpful to refer to Figure 8-10.

Let k be the largest index on $0, \ldots, m$ such that, for every vertex v_j on the path with $j \le k$, v_j (or a blossom that v_j lies in within the collapsed graph) is in the search tree rooted at v_0 (or rooted at a blossom containing v_0). Observe that $k \le m - 1$, v_k

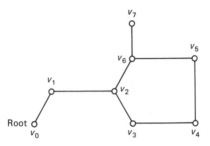

Figure 8-10. Diagram for proof of correctness.

must be an inner vertex of the search tree, and the vertex matched with v_k must be v_{k-1}, from which it follows that the index k must be even.

Now, let r be the smallest even index such that v_r is inner. By our initial observation, one of the vertices $v_0, \ldots, v_{r-1}$ must be inner, or else $v_0, \ldots, v_r$ would have been collapsed into a single blossom by the time the search tree was blocked, contrary to v_r being inner. Therefore, we can let i be the largest index less than r such that v_i is inner. Then, i must be odd, and so the matched vertex to v_i must be v_{i+1}. Therefore, the vertices $v_{i+1}, \ldots, v_{r-1}$ must be collapsed to a single vertex which is matched to both v_i and v_r, which is a contradiction. It follows that Find_Augmenting_Path must find some augmenting path, if one exists, which completes the proof of the theorem.

REFERENCES AND FURTHER READING

Edmonds' famous algorithm was given in Edmonds (1965). There have been a variety of improvements since then, including improved algorithms for bipartite graphs. Gabow (1976) and Lawler (1976) show how to implement Edmonds' algorithm in $O(|V|^3)$ time. A special case linear time disjoint set union algorithm of Gabow and Tarjan (1983) yields an algorithm with time $O(|E||V|)$. See also Tarjan (1983), which, along with Edmonds' original article, is the basis for some of our discussion. For an extensive discussion of results and algorithms on matching, see Swamy and Thulasiraman (1981). Behzad, et al. (1979) also gives the basic theory of covers, etc. The proof of Gallai's Theorem is based on Behzad, et al. (1979). Valiant (1975) gives the optimal parallel selection analysis. Refer to Turan (1954) for a proof of the extremal theorem Valiant uses. See Coffman and Denning (1973) for a discussion of the two-processor scheduling algorithm. The deadlock-free system characterization and algorithm is after Kameda (1980). The reducing path characterization of a minimum edge cover in the exercises is from Norman and Rabin (1959).

EXERCISES

1. Prove that a tree has at most one perfect matching.
2. Use the duality implied in Gallai's Theorem to show how to construct a minimum edge cover from a maximum matching.
3. An *alternating path with respect to an edge cover* C on a graph $G(V, E)$ is a path whose edges are alternately in C and outside C. A *reducing path with respect to an edge cover* C is an alternating path whose terminal edges are in C and whose terminal vertices are incident to edges of C which are not terminal edges of C. Prove the following analog of the augmenting path characterization for maximum matchings. An edge cover C is a minimum cover if

and only if there is no reducing path with respect to C. For this condition to work, must we allow the reducing path to be closed? Give an example, or clearly address this in the proof.

4. What is the relationship between a coloring of a graph and a partition of a graph into independent sets of vertices?

5. Give an algorithm that finds a maximum cardinality vertex independent set on a bipartite graph.

6. Consider the following method for finding a maximum independent set in a graph $G(V, E)$. Let v be a vertex in G of degree at least 3. Either v is in a maximum independent set of G or it is not. If it is, then none of the vertices in ADJ(v) can be in the maximum independent set. If it is not, then the maximum independent set is contained in $G - \{v\}$. Develop these observations into a recursive algorithm for finding a maximum independent set, and determine a bound on the performance of the algorithm using recurrence relations. Since this problem is NP-Complete, the bound is exponential.

7. Prove that every bipartite graph G is a subgraph of a max(G)-regular bipartite graph.

8. Prove there are at most $3^{(|V|/3)}$ maximal independent sets in a graph $G(V, E)$.

9. Prove that if VC(G) $= k$, then at least one of the colors in a k-coloring of G determines a maximal independent set.

10. Prove that every cubic graph with no articulation point is 1-factorable.

11. Construct a cubic graph containing no 1-factors.

12. Use the maximum-flow minimum-cut theorem to show that if G is bipartite, Max-Match(G) equals Min-VCov(G).

13. Model the degree constrained maximum matching problem on a bigraph as a flow problem.

14. Write a program that determines whether a system is deadlock free, or if it is not, outputs a resource request/allocation sequence leading to a deadlock.

15. Let $G(V, E)$ be a bipartite graph, and let M be a matching on G. Then, prove there is a maximum matching M' on G such that every vertex matched in M is also matched in M'.

16. Let $T(V, E)$ be a tree, and let o(T, v) be the number of components of $T - v$ of odd order. Prove that T has a perfect matching if and only if o(T, v) equals 1 for every v in T.

17. Construct a counterexample to the following greedy algorithm for the minimum vertex cover of a graph G, and try to estimate its worst case error (assume G has no isolated vertices). Repeat the following until G is empty: Select a vertex v of maximum degree; add v to the cover; and delete its incident edges and any resulting isolated vertices from G.

18. A vertex independent set in $G(V, E)$ corresponds to a complete subgraph in the complement of G. Referring to the exercise on a backtracking algorithm for cliques in Chapter 2, show how this relates to a backtracking algorithm for independent sets.

19. Show how to find the maximum number of vertex disjoint paths between a given pair of sets of vertices X and Y in a graph $G(V, E)$ by modelling the problem as a matching problem. (See the section on problem transformation in Chapter 2, where this is done for the case where X and Y have the same cardinality.)

9

Parallel Algorithms

The architectural, communications, scheduling, and synchronization issues that arise in parallel algorithms add a new dimension to the design and analysis of algorithms. This section describes parallel algorithms both for some highly specific parallel computer architectures, like the systolic array and the tree processor, as well as the more loosely specified shared memory model of parallel computation. The kinds of synchronization used for these algorithms range from the rigid lock-step synchronization characteristic of the systolic architecture, to the explicit use of synchronization primitives for a parallel system of processors for updating a heap, to shared memory algorithms where synchronization is largely implicit. Performance measures for parallel algorithms, and their basic complexity classes, are presented in Chapter 10.

9-1 SYSTOLIC ARRAY FOR TRANSITIVE CLOSURE

We will describe a systolic processor array that computes matrix products and can be used to compute the transitive closure of a digraph. A *systolic processor array* is a collection of rudimentary processors, each capable of performing some simple operation, interconnected by a fixed and regular network, and operating in a highly synchronous fashion. The operation of a systolic array is determined by

(1) The function of its individual processing elements,

(2) The pattern of processor interconnection, and

(3) The arrival order pattern for the input.

The function of the processing element (PE) depends on the application. We use the element shown in Figure 9-1. Its function is defined in terms of the effects it has on its inputs. In this case, the element has three inputs and three outputs. The inputs are numeric operands a, b, and c. These are read at the beginning of the processing cycle at the ports shown in Figure 9-1a. Then, at the end of the cycle, the PE outputs a and b without alteration, computes ab, and outputs $c + ab$, as shown in Figure 9-1b.

Depending on the processing to be accomplished, different interconnection patterns between the PEs are used. Some common patterns are linear, rectangular, and hexagonal arrays. Our application uses the rectangular pattern shown in Figure 9-2.

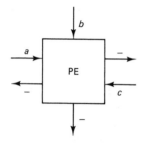

(a) Systolic processing element PE with inputs.

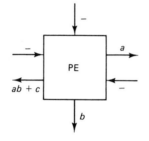

(b) Outputs of PE after systolic operation.

Figure 9-1. Function of systolic processing element.

Data flow. The proper presentation of inputs to the processor is critical. In our problem, the array inputs a pair of n by n matrices **A** and **B**, and outputs the product **C** equal to **AB**. The inputs are presented as shown in Figure 9-2. **A** arrives at the left border of the systolic array and **B** enters in a skewed manner along the top border. This arrangement ensures the inputs interact correctly.

The ith row of **A** is fed to the ith row of the systolic array. Its components are interspaced within zeros, which provide the delay needed so the components of **A** and **B**

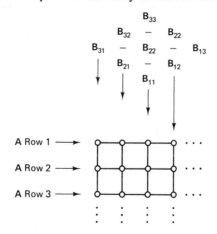

Arrival order for A

$\cdots$	$\cdots$	$\cdots$	0	A_{13}	0	A_{12}	0	A_{11}
$\cdots$	$\cdots$	0	A_{23}	0	A_{22}	0	A_{21}	0
$\cdots$	0	A_{33}	0	A_{32}	0	A_{31}	0	0

Arrival order for B

$\cdots$	0	B_{42}	0	B_{33}	0	B_{24}	$\cdots$
$\cdots$	B_{41}	0	B_{32}	0	B_{23}	0	$\cdots$
$\cdots$	0	B_{31}	0	B_{22}	0	B_{13}	$\cdots$
$\cdots$	0	0	B_{21}	0	B_{12}	0	$\cdots$
$\cdots$	0	0	0	B_{11}	0	0	$\cdots$
$\cdots$	0	0	0	0	0	0	$\cdots$

Figure 9-2. Arrival order for inputs to AB systolic processor.

can interact requisitely. For the same reason, the successive **A** rows are input with increasing stagger delays. This gives the **B** components, which are fed into the systolic array processor from the upper border, time to move through the array to the lower rows where they can interact with their **A** counterparts.

The arrival pattern for **A** determines the arrival pattern for **B**. For example, in order for the first row of **A** to meet its counterpart in **B**, we need to input the ith column of **B** in a skewed manner, and interspaced with dummy zero values. **B** can be visualized as being tilted, that is, transposed and rotated through 135 degrees, with $\mathbf{B(1, 1)}$ entering at the midpoint of the upper border of the systolic array, and the rest of the components falling into place accordingly. Referring to Figure 9-2, $\mathbf{B(1, 1)}$ enters the array at column n; $\mathbf{B(1, j)}$ enters at column $n + (j - 1)$ and $\mathbf{B(i, 1)}$ enters at $n - (i - 1)$. In general, $\mathbf{B(i, j)}$ enters at column $n + (j - 1) - (i - 1)$.

The input presentation can be specified more formally, and there are even techniques for systematically deriving the systolic architecture and data presentation appropriate for a particular combinatorial problem. We refer the reader to the references. The following theorem establishes the correctness and performance of the system we have described.

Theorem (Correctness and Performance of Systolic Matrix Multiplication Algorithm). Let **S** be the systolic array defined above, and suppose **A** and **B** are input as indicated. Then, the system correctly computes the components of the product **C** equal to AB. The component $\mathbf{C(i, j)}$ is output from the leftmost PE in row i of S at time $2n + i + 2j - 3$.

The proof of the theorem is as follows. We first prove that the algorithm correctly computes AB, by showing that the right matrix components of **A** and **B** arrive at the right processors at the right points in time. Recall that each component $\mathbf{C(i, j)}$ of the product is a sum of terms of the form $\mathbf{A(i, k)B(k, j)}$. Thus, we must prove the components $\mathbf{A(i, k)}$ and $\mathbf{B(k, j)}$ meet at a common PE at a common point in time, as well as that the successive terms of the sum defining $\mathbf{C(i, j)}$ are accumulated correctly.

Refer to Figure 9-3 for the following. Consider a pair of components $\mathbf{A(i, k)}$ and $\mathbf{B(k, j)}$. These move across the ith row of S and down the $(n + (j - 1) - (k - 1))$-th column of S, respectively. Their arrival times at the various PEs depend on several factors: the delay involved in the data moving through S (the *systolic data flow delay*), the delay before the input rows even arrive at the input ports of S (the *input stagger delay*), and the delay due to the spacing of input elements with dummy values (*the spacing delay*). Obviously, the stagger and spacing delays are chosen so the two input data flows, from **A** and from **B**, interact correctly. With the choices of delays indicated in Figure 9-3, both $\mathbf{A(i, k)}$ and $\mathbf{B(k, j)}$ eventually pass through the PE $P(i, n + (j - 1) - (k - 1))$ at time $n + (i - 1) + (k - 1) + (j - 1)$. That element then computes their product, accumulates it into a partial sum received from the PE to its right, and then passes the sum on to the processing element $P(i, n + (j - 1) - k)$ that lies to its left, where its value is then available at the next systolic pulse at time $n + (i - 1) + (k - 1) + (j - 1) + 1$.

To prove that the partial sums are accumulated correctly, we argue as follows. Consider the continuation of the calculation of the partial sum for $\mathbf{C(i, j)}$ at the next PE. Observe that the processing element $P(i, n + (j - 1) - k)$ on the left side of $P(i, n +$

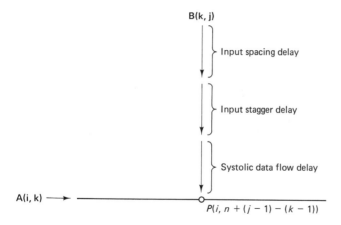

B(k, j)

Input spacing delay

Input stagger delay

Systolic data flow delay

A(i, k) $\longrightarrow$

$P(i, n + (j - 1) - (k - 1))$

Arrival at $P(i, n + (j - 1) - (k - 1))$:

Delay: Spacing-delay + stagger-delay + systolic-delay

A(i, k): $2(k - 1) + (i - 1) + (n + (j - 1) - (k - 1))$

B(k, j): $((k - 1) + (j - 1)) + (n - 1) + i$

Figure 9-3. Data timing.

$(j - 1) - (k - 1))$ must receive the components $A(i, k+1)$ and $B(k+1, j)$ at the next time pulse. But, the component $A(i, k+1)$ is scheduled to arrive at this processor at time $(i - 1) + (2k) + (n + (j - 1) - k))$ or equivalently at time $n + (i - 1) + (j - 1) + (k - 1) + 1$, which is the same as the time at which the component $B(k+1, j)$ arrives, as required. It follows that the systolic array correctly accumulates the needed partial sums.

The performance of the algorithm follows directly from the previous timing considerations. The last element computed for $C(i, j)$ is the product $A(i, n)B(n, j)$ which as we have seen is computed at PE $P(i, j)$ at time $n + (i - 1) + (j - 1) + (n - 1)$. Taking into account the systolic data flow delay, this output arrives at the output port on the ith row of the array at time $2n + i + 2j - 3$. The first component is thus output at time $2n$. The very last component $C(n, n)$ is output at time $5n - 3$. This completes the proof of the theorem.

Thus $O(n^2)$ systolic PEs compute a matrix product in $O(n)$ time, for a total space-time cost of $O(n^3)$. In contrast, the corresponding sequential algorithm uses a single processor and takes $O(n^3)$ time, and the same space-time cost. Refer to Figure 9-4 for a detailed example.

Systolic transitive closure. The matrix product systolic array can be adapted to compute the transitive closure of a digraph. Recall that if A is the adjacency matrix of a graph $G(V, E)$, then the (i, j) component of A^k equals the number of walks from vertex i to j with k edges. Thus, the (i, j) component of $(I + A)^{|V|-1}$ is nonzero if and only if there exists a path from i to j, whence we can use it to determine the transitive closure of G.

This computation can be interpreted from a boolean viewpoint. Thus, define A to be a *boolean adjacency matrix* for a digraph if $A(i, j)$ is true, if (i, j) is an edge or i

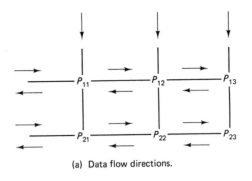

(a) Data flow directions.

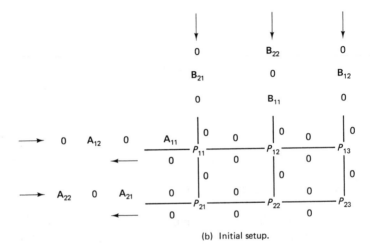

(b) Initial setup.

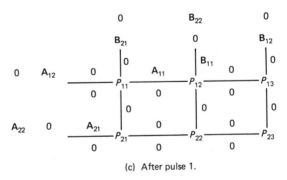

(c) After pulse 1.

Figure 9-4. Trace of operation of systolic array for AB.

equals j, and false otherwise. The (i,j) component of the product of a pair of Boolean matrices $\mathbf{A}$ and $\mathbf{B}$ is

$$\bigcup_{k=1}^{|V|} (\mathbf{A_{ik}} \cap \mathbf{B_{kj}}) .$$

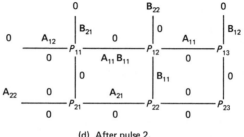

(d) After pulse 2.

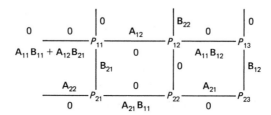

(e) After pulse 3: C_{11} computed and available.

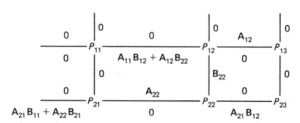

(f) After pulse 4: C_{21} available, C_{12} computed.

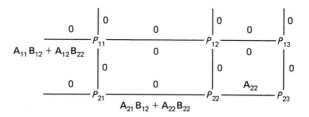

(g) After pulse 5: C_{12} available, C_{22} computed.

(**Figure 9-4** continued)

The components of the boolean square of **A** indicate whether or not there is a walk of length 2 or less from i to j, while in general, the boolean power $A^{|V|-1}$ equals the boolean adjacency matrix for the transitive closure of G. If $|V| - 1$ equals 2^k, then $\mathbf{A}^{|V|-1}$ can be found in k stages using the product systolic processor, adapted to perform boolean multiplication, to iteratively compute: $\mathbf{A}^2$, $(\mathbf{A}^2)^2$, $(\mathbf{A}^4)^2$, and so on. This also holds with minor modification for arbitrary values of $|V|$. Therefore, we can compute the transitive closure of a digraph in $O(|V| \log |V|)$ steps using a systolic array.

9-2 SHARED MEMORY ALGORITHMS

Much of the theoretical research on parallel processing has used a very unrestrictive model of parallelism called the *Shared Memory* (SM) model. In this model, a pool of synchronized parallel processors share access to a common global memory. It is customary to classify algorithms for this model according to the type of memory access they require. For example, in one SM parallel version of Dijkstra's algorithm, none of the parallel processors either reads from or writes to the same location concurrently, so the algorithm requires only a memory supporting this type of access.

It is customary to distinguish four different kinds of access.

(1) EREW (Exclusive Read/Exclusive Write)

(2) CREW (Concurrent Read/Exclusive Write)

(3) CRCW (Concurrent Read/Concurrent Write)

(4) ERCW (Exclusive Read/Concurrent Write)

For example, an EREW algorithm is one which makes no concurrent reads or writes to a given memory location. This is the case for the parallel Dijkstra's algorithm. A CREW algorithm makes concurrent reads to a given memory location, but no concurrent writes to a given location. This is the case for the parallel Floyd's algorithm we will describe. A CRCW algorithm makes both concurrent reads and writes to given memory locations. In the case of a concurrent write, only one of the writes is successful. This is the model used for a connected components algorithm we will describe.

The shared memory is referred to as a *Parallel Random Access Memory* or PRAM. Algorithms are then called EREW PRAM, etc., according to their memory access requirements. We generally ignore how the memory requirements are implemented architecturally. However, any of the various SM models can be simulated using an interconnection network that links the pool of parallel processors and their common memory. But, for a system with n parallel processors, such an implementation slows memory access over the idealized SM model by a logarithmic factor. We refer to Ullman (1984) for a detailed discussion of the simulation of idealized parallel computers and for further references, and to Vishkin (1983).

9-2-1 Parallel Dijkstra Shortest Path Algorithm (EREW)

We have previously described Dijkstra's shortest path algorithm (Chapter 3). We will now show how to implement it in parallel using an EREW PRAM with p processors. The results are summarized in the following theorem.

Theorem (Parallel Dijkstra Performance). Let $G(V, E)$ be a network with nonnegative edge weights, and let x and y be a pair of vertices in G. Then, it is possible to find the shortest distance between x and y using a parallel version of Dijkstra's algorithm on an EREW PRAM with $p \leq |V|$ processors in $O(|V|^2 / p + |V| \log p)$ time. For $p \geq |V| / \log |V|$, this is $O(|V| \log |V|)$.

To prove the theorem, we need to establish bounds for the following two steps, which are the performance bottlenecks in Dijkstra's algorithm:

(1) Find the next vertex v to include in the shortest path tree, where v is the search tree vertex with the smallest estimated distance from the start vertex, and which is not yet in the shortest path tree.

(2) Extend the search tree from v to its neighbors by updating the estimated distances of the neighbors.

The following lemmas will be helpful.

Lemma 1. Let S be a set of $|V|$ numbers. Suppose there are $p \leq |V| / 2$ processors available which utilize an EREW PRAM. Then, the minimum value of S can be found in $O(|V| / p + \log p)$ time.

The proof of lemma 1 utilizes a simple parallel divide and conquer technique. We consider first the case where $p = |V| / 2$ processors are available. For simplicity and without loss of generality, we assume $|V|$ is a power of 2. We organize the required computations in the fashion of a binary tree. At the level of the endpoints of the tree, we group the values of S into $|V| / 2$ pairs, find the minimum of each of the pairs using $|V| / 2$ processors (in constant time), then group the resulting $|V| / 2$ minima into $|V| / 4$ new pairs. We then repeat the process, finding the minimum of each of the $|V| / 4$ pairs in constant time, and reusing $|V| / 4$ of the original p processors for this round of computation. Each time the process is repeated, the number of processors required as well as the number of pairs of values is halved until after $\log |V|$ stages we have found the minimum of S. A statement of this procedure follows, where $|V|$ equals 2^n:

Parallel_Procedure Minimum (M)

(* Returns the minimum of M = [M$_1$, ..., M$_{|V|}$] in M$_1$ using
 $|V|/2$ processors on an EREW PRAM *)

var |V|: Integer constant
  M(|V|): Real
  n: Integer
  i: 0..n − 1
  j: 1..|V| − 1

Set n to lg (|V|)

for i = 0 to n − 1 **do**

  for j = 1 to |V| − 1 in steps of 2^{i+1}

    do_in_parallel Set M$_j$ to min { M$_j$, M$_{(j + 2i)}$ }

End_Parallel_Procedure_Minimum

When $p < |V| / 2$ processors are available, we use a combination of sequential and parallel processing to obtain an improvement in performance proportional to p. The idea is to partition the $|V|$ values into p groups of size $|V| / p$ each. We then assign a single processor to each group which finds the smallest element of the group in sequential time $O(|V| / p)$. Then, we apply the parallel method to find the smallest of these p minima. This takes time $O(\log p)$ using p processors, for an overall performance of $O(|V| / p + \log p)$, which completes the proof of lemma 1.

Lemma 2. Assume $p \leq |V| / 2$. Then, step (2) of Dijkstra's algorithm can be done on a p processor EREW PRAM in $O(|V| / p + \log p)$ time.

The proof is as follows. We first broadcast the identity of v and its distance from the root to the other processors. This can be done in $O(|V| / p + \log p))$ time using a binary tree broadcasting scheme (that is, one processor tells another processor, then they each tell two other processors, and so on. As in the preceding lemma, the last broadcasting step may be done sequentially/in-parallel if there are $< |V| / 2$ processors.) We then update the estimated distances to each of the outdeg(v) neighboring vertices in parallel in $O(\text{outdeg}(v) / p)$ steps, which is dominated $O(|V| / p)$. This completes the proof of lemma 2.

The proof of the theorem now follows readily. Recall that for each phase of Dijkstra's algorithm there are at most $O(|V|)$ vertices considered for inclusion in the search tree, the one with the smallest estimated distance being selected. By lemma 1, we can find this vertex using $p \leq |V| / 2$ processors in $O(|V| / p + \log p)$ time. If we sum the performance estimates from lemmas 1 and 2 over the at most $|V|$ phases of the algorithm, the total time, for finding minima, broadcasting, and revising values, is

$$O(|V|^2 / p + |V| \log p)) .$$

For $p \geq |V| / \log |V|$, this is just $O(|V| \log |V|)$. This completes the proof of the theorem.

9-2-2 Parallel Floyd Shortest Paths Algorithm (CREW)

The sequential version of Floyd's shortest paths algorithm for a weighted digraph $G(V, E)$ can be almost directly reinterpreted as a parallel algorithm on a CREW PRAM. We assume G is represented in the same manner as for the sequential version of Floyd's algorithm, that is, as a packaged distance matrix, *Dist*. The parallel algorithm follows.

```
Parallel Procedure Floyd (G,SD)

(* Returns the matrix of shortest distances on G in SD *)

var  G: Graph
     i,j,k: 1..|V|
     SD(1..|V|, 1..|V|): Real

for i, j = 1..|V| do_in_parallel   Set SD(i,j) to Dist(i,j)

for k = 1 to |V| do

   for i, j = 1 to |V| do_in_parallel
       Set SD(i,j) to min {SD(i,j), SD(i,k) + SD(k,j)}

End_Parallel_Procedure_Floyd
```

The validity and performance of the algorithm are summarized in the following theorem.

Theorem (Parallel Floyd). Let $G(V,E)$ be a network with real valued edge weights. Then, a parallel version of Floyd's algorithm can be implemented on a CREW PRAM with $p \leq |V|$ processors in $O(|V|^3 / p)$ time.

The proof is as follows. Since we already know the sequential algorithm is correct, we only need to prove that the parallel computations yield the same results. Referring to Figure 9-5, recall that every entry in SD that can change at stage k depends only on its current value and the values of a pair of components in the kth row and column. As was observed for the sequential algorithm, neither of these can change at stage k. Consequently, the components of SD may be updated in parallel, using only EW access to memory, since no memory location is written to by more than one processor. The algorithm does, however, require CR access, since the components of the kth row and column are read concurrently, by as many as p processors. When $p = |V|^2$ processors are available, each stage takes $O(1)$ time, so the algorithm takes $O(|V|)$ time. In general, if $p \leq |V|^2$ processors are available, the algorithm is slowed up proportionately, so the time becomes $O(|V|^3 / p)$. This completes the proof of the theorem.

9-2-3 Parallel Connected Components Algorithm (CRCW)

We now describe a CRCW algorithm that computes the components of a graph $G(V,E)$ using $|V| + 2|E|$ processors (one per vertex and two per edge) in $O(\log |V|)$ time. The algorithm uses a simple component merging technique, where vertices are initially assigned to their own components, which are then merged as edges between components are recognized, until eventually the true components are found. The technique is complicated only by the CRCW computing environment.

We denote the vertex processors by Vertex_processor(i), $i \leq |V|$. $G(V,E)$ is represented by an edge list where each undirected edge is represented *twice*, once under (i,j) and once under (j,i). Edge_processor(i,j) handles the processing for representative (i,j). The processors are synchronized, beginning each step at the same time, and writing simultaneously. If any processors concurrently write to a common location, an indeterminate one of the writes is successful.

The algorithm uses a unary tree representation of the components (refer to Chapter 4.) The parent of a tree vertex v is denoted by Par(v), and a tree root r is distin-

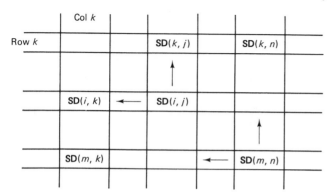

Figure 9-5. Data dependencies for Floyd's algorithm.

guished by the fact that r equals Par(r). A pair of vertices lie in the same component if their Par pointer trails lead to the same root. The unary trees initially consist of single vertices. The basic operations on the unary trees are "shorten" and "merge" (or "hook"). Shorten is implemented by setting Par(v) to Par(Par(v)), for each vertex v at regular intervals; while we hook a vertex v to Par(w) by setting Par(Par(v)) to Par(w). The vertex processors implement the shorten operations, and the edge processors implement the merge operations. The shorten operation reduces the height of the trees, thus keeping the Par pointer trails short. By termination, the trees are all stars (of unit height), so component membership can be tested in $O(1)$ time.

The two hooking operations (used in steps [2] and [3] of the algorithm, respectively) merge unary trees, and thus the components they represent, by making a root of one tree point to a vertex of the other. The step (2) operation hooks a root of one tree to a vertex of lower index in another tree. That is, each Edge-processor(i, j) whose vertex i pointed to a root at the end of the previous iteration, "tries" to set Par(Par(i)) to Par(j) if that has a smaller index. The word "tries" is appropriate because the attempted write may fail because of a competing concurrent write! The step (3) operation uses a field Flag(i) at each vertex i which is turned off at the beginning of each iteration, and turned on only if i is involved in a shorten operation in step (1) or a hook operation in step (2). The operation applies only to a vertex i which is a root or a child of a root whose flag is unset. The operation hooks the root to a vertex in another tree (whose index may not be lower).

The procedure definition follows. It uses a function Test(Flag), which succeeds if Flag(v) has been set for any v, and fails otherwise. Figure 9-6 gives an example: The two parallel columns show the different outcomes that occur depending on whether v_4 is initially linked to v_3 or v_1. Either outcome is possible because of the indeterminate nature of the concurrent writes.

Parallel_Procedure Components (G, Par)

(* Returns in Par(i) the component number of vertex i *)

var G: Graph
Par(|V|), Par'(|V|), i, j: 1..|V|
Flag(|V|): 0..1
Test: Boolean function

Initialize the Vertex and Edge processors using $G(V,E)$

for every Vertex-processor(i)

   do_in_parallel **Set** Par'(i) to i
          Set Par(i) to i

repeat

(1) Shorten the trees:

   for every Vertex-processor(i) **do_in_parallel**

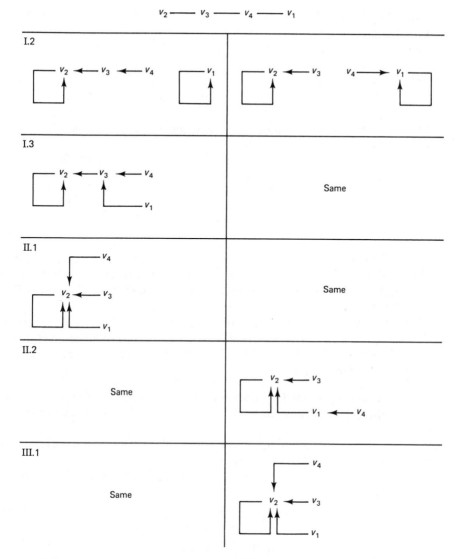

Figure 9-6. Column 1 (Column 2) shows the outcome if v_4 is initially linked to v_3 (v_1).

          Set   Flag(i) to 0; **Set**   Par(i) to Par'(Par'(i))
          if   Par(i) <> Par'(i)   **then**   **Set** Flag(Par(i)) to 1

    (2)   Merge components by hooking roots to lower indexed roots:

          for every Edge_processor (i,j) **do_in_parallel**

          if   Par(i) = Par'(i) (ie, Par(i) was already a root)
          then **if**   Par(j) < Par(i)
                  then Set Par(Par(i)) to Par(j)
                     Set Flag(Par(j)) to 1

(3) Merge components by hooking unflagged roots to other trees:

    for every Edge_processor (i,j) **do_in_parallel**

        if   Par(i) = Par(Par(i)) **and** Flag(Par(i)) <> 1
        then **if** Par(j) <> Par(i)
             then Set Par(Par(i)) to Par(j)

(4)   **for** every Vertex-processor (i) **do_in_parallel**

        Set Par(i) to Par(Par(i)); **Set** Par'(i) to Par(i)

    until **not** Test(Flag)

    End_Parallel_Procedure_ Components

Before proving the correctness and performance of the algorithm, we need to introduce some notation. Let F denote the *pointer graph* determined by the Par pointers. (Later we will see F is a forest of unary trees.) For v in $V(F)$, let $S(v)$ denote the subgraph induced in F by the set of vertices in F that can reach v via a path in F. The *order* of $S(v)$ is denoted by $C(v)$, and the length of the longest path in $S(v)$ ending at v (or, the *height* of v in $S(v)$) is denoted by $H(v)$. Let $F(s, k)$ denote *the pointer graph after the execution of step k ($k = 1, \ldots, 4$) of iteration s.* If v is a root and Flag(v) is unset after steps (1) and (2) of an iteration, then v is called a *stagnant root*. Our results are summarized in the following theorem.

    Theorem (Connected Components Correctness).   Let $G(V, E)$ be a graph. Then, the parallel procedure Components correctly computes the components of G in $O(\log |V|)$ time.

    We prove the theorem using a sequence of lemmas, the first several of which establish properties of the graphs $F(s, k)$.

    Lemma 1. If v is a stagnant root after step (2) of any stage s, then $H(v)$ is ≤ 1 in $F(s, 1)$, $F(s, 2)$, and $F(s, 3)$.

    To prove this, observe that if v is stagnant after step (2), then $H(v)$ must be ≤ 1 in step (1); otherwise $S(v)$ would have been shortened in step (1) and so v would not be stagnant. $H(v)$ cannot then become greater than 1 in step (2), since for that to happen a root would have to be hooked onto v in step (2), again contrary to v being stagnant after (2). For $H(v)$ to be greater than 1 in step (3), would again require hooking another necessarily stagnant root w onto v in that step. But, it is straightforward that if $v = \text{Par}(i)$ and $w = \text{Par}(j)$ are distinct stagnant roots after step (2), then (i, j) cannot be an edge. Therefore, no stagnant root w can be hooked onto v, and so $H(v)$ must remain ≤ 1 in step (3). This completes the proof of lemma 1.

    Lemma 2. If the index of a vertex is less than the index of its parent (that is, $i < \text{Par}(i)$) in $F(s, k)$ ($k = 1, 2,$ or 4), then i must be an endpoint of $F(s, k)$.

    The proof is by induction (over the lexicographically ordered tuples (s, k)). The result holds trivially for step (1) of stage 1. Therefore, we start with step (2) of stage 1. If the lemma holds after step (1) of any stage, then it holds after step (2) of that stage because we never hook a vertex to a larger indexed vertex in step (2). On the other hand, it can happen that in step (3) we hook a vertex i onto a larger indexed vertex; but then i must have been stagnant at the beginning of step (3) and so by our previous observations $H(i)$ must have been ≤ 1 at that point. Therefore, by the end of the next step (4), i will once again be an endpoint, as required. Finally, the step (1) operation of the next stage only creates new endpoints, and destroys none of the existing endpoints; so the result still holds. This completes the inductive proof of this result.

Lemma 3. $F(s, k)$ is a forest, with self-loops at the roots of its component trees.

To prove this, first observe that step (1) never introduces a cycle. If a cycle were caused in step (2), let i be the largest indexed vertex on that cycle. The cycle predecessor of i would have an index less than i, contrary to lemma 2. Finally, step (3) only hooks stagnant roots to nonstagnant roots; so in no case do we ever hook a tree to itself. Thus, $F(s, k)$ is a forest.

Lemma 4. If the algorithm terminates, then each component of F at termination is a star with vertices equal to the vertices of a component of G.

The proof is as follows. If v is a root which is stagnant after step (2) of any stage s, and which remains a root after step (4) of that same stage, then the component $S(v)$ of $F(s, 2)$ containing v must be a star, and every vertex in the same component of $G(V,E)$ as v must be already included in $S(v)$. Otherwise, let w be the nearest vertex reachable from v in G which is not in $S(v)$. In step (3) of stage s, v would have been hooked either to Par(w) or to another vertex not yet in $S(v)$, contrary to the supposition that v is still a root after step (4). It is also obvious that $S(v)$ never changes thereafter.

Lemma 4 proves the algorithm is correct, provided it terminates. Furthermore, at termination, we can test pairs of vertices for membership in a common component in constant time, because of the star-like character of the final components of F: we just test whether the vertices point to a common parent. It remains to prove the algorithm terminates. The following lemma is critical.

Lemma 5. If v is a root of any tree in F that has changed during the s-th iteration, then $C(v) \geq H(v) (3/2)^{s-1}$.

The proof follows. The initial inequality is trivial. Generally, observe that if a tree does not change during an entire iteration, it must already be in its final form, and so will not change in any subsequent iteration. Consequently, we can assume that if the tree containing v changes during iteration s, the tree must also have changed during iteration $s - 1$. Therefore, we may assume by induction that $C(v) \geq H(v) (3/2)^{s-2}$. We distinguish two cases, according to whether v was the root of a star before step (4) of iteration s, or was not.

Case (1): If v was the root of a star before step (4) of iteration s, then v must have been the root of the very same star at the end of step (1), because after step (1) all the trees (for nontrivial components) are already nontrivial, and so the hooking operations in steps (2) and (3) could never yield a star. Thus, the star must not have changed in either steps (2) or (3). Consequently, the star must have changed (by shortening) during step (1), and so was reduced in height by at least a factor of 3/2, as required. (Observe that the height reduction accomplished by shortening depends strongly on the acyclic nature of F. If F had cycles, then not only might no decrease in height occur from a shortening, but the procedure could loop endlessly.)

Case (2): If v was not the root of a star before step (4) of iteration s, then observe that if the algorithm merges in this step trees T_i in F with roots v_i and heights h_i, $i =$

$1, \ldots, n$, the resulting tree has height at most $h_1 + \ldots + h_n$. (Remember that by the parallel nature of the algorithm, it can merge many of these trees concurrently! For example, it might hook v_1 to v_2 while it is at the same time hooking v_3 onto v_1.) It follows by induction that after step (3) of stage s, the inequality $C(v) \geq H(v)(3/2)^{s-1}$ still holds, since the subtree cardinalities combine additively, while the heights combine subadditively. The shortening in step (4) then decreases the heights by a factor of $3/2$, as required. This completes the proof of the lemma.

The proof of termination is now straightforward. For, since the order of G is fixed, a logarithmic bound on the total number of stages follows easily from the exponential lower bound on the size of the still varying components of F guaranteed by lemma 5. Indeed, it readily follows that the algorithm must terminate in $O(\log |V|)$ stages, since otherwise some $C(v)$ would be greater than $|V|$. This completes the proof of the theorem.

9-2-4 Parallel Maximum Matching Using Isolation (CREW)

If a set of parallel processors are used to construct some optimal combinatorial object, how do we ensure that they all construct the same object? One approach is to have all the processors construct an object with a unique property, for example, the lexicographically smallest optimal object. But, this problem is often NP-Complete. An alternative is to randomly assign weights to the elements of a problem, so the optimal object becomes a weighted optimal object which is unique with some fixed probability. This enables the parallel processors to operate consistently. The technique is called *isolation*. We will illustrate it for the problem of matching.

For the problem of parallel matching, isolation is done as follows. We randomly assign weights to the edges of a graph $G(V, E)$ so that G contains, with some nonzero probability, a unique minimum weight perfect matching. We then define a property which is a function of the edge weights, but is satisfied only by edges in the unique optimal matching, and can be tested in parallel for each edge, allowing us to design a parallel matching algorithm. Because of its probabilistic character, the isolation technique may fail to isolate a unique solution. Therefore, we must be able to detect when failure occurs and recover from it. For the problem of matching, failure is easily detected: The output is not a perfect matching. The recovery procedure is simply to try again. If each attempt has a fixed probability, say, $1 - p$ of success, then the probability p^k of k failures decreases to zero rapidly.

Let us define a *family of sets* (S, F) as a set of elements $S = \{x_1, \ldots, x_n\}$ together with a family F of subsets of S. If $G(V, E)$ is a graph, taking S equal to $E(G)$ and F equal to the set of perfect matchings on G is an example of a family of sets. The following theorem shows that, given an arbitrary family of sets, we can randomly assign weights to the elements of S, so that with high probability we determine a unique set in F.

Theorem (Isolating Lemma). Let S be a set of size n, and let (S, F) be a family of sets on S. Assign integer weights chosen uniformly and independently from $[1, 2n]$ to each element of S. Then, the probability that there is a unique set in F of minimum weight is at least $1/2$.

The proof is as follows. For each element x_i in S, fix the weights of the remaining elements of S. Let $Low(x_i)$ denote the unique real number such that if $w(x_i) \leq Low(x_i)$, then x_i is a member of some minimum weight set in F, while if $w(x_i) > Low(x_i)$, then x_i is not a member of any minimum set in F. It follows that if $w(x_i) < Low(x_i)$, then x_i belongs to every minimum subset; while if $w(x_i) = Low(x_i)$, then some minimum sets contain x_i and some do not (in which case we call x_i *singular*).

The actual weight w_i assigned to x_i is independent of $Low(x_i)$ (which is only a function of (S,f), x_i, and the weights other than w_i) and is uniformly distributed on $[1, 2n]$. Therefore, the Prob $\{x_i$ is singular$\}$ (= Prob $\{ w_i = Low(x_i)\}$) is independent of i; hence equal for every i, and so at most equal to $1/(2n)$. It follows that Prob {some element of S is singular} is at most $n/(2n)$ or $1/2$; or equivalently, the probability that no element of S is singular or that F has a unique minimum weight subset, is at least $1/2$. This completes the proof of the theorem.

The procedure of the theorem isolates a unique set in F at least half the time. If we can detect when isolation fails, then we can, on failure, repeat the procedure with a new random assignment of weights with the probability of k successive failures being at most $(1/2)^k$, whence the probability of success can be made arbitrarily high. We apply the theorem to the perfect matching problem by assigning random weights from $1 .. 2|E|$ to the edges; so a (randomly) unique minimum weight perfect matching becomes the well-defined object of the parallel search.

Perfect bipartite matching. Let $G(U, V, E)$ be a bipartite graph with a perfect matching. Denote the vertices in U by $\{u_1, \ldots, u_n\}$ and those in V by $\{v_1, \ldots, v_n\}$.

Problem: Perfect Bipartite Matching
Input: A bipartite graph $G(U, V, E)$ with a perfect matching
Output: A perfect matching on G

We can solve this problem using the Isolating Lemma. Denote the *bipartite adjacency matrix* of G by $\mathbf{A}$, where $\mathbf{A}$ is the n by n matrix such that $\mathbf{A}(\mathbf{u_i}, \mathbf{v_j})$ is 1 if (u_i, v_j) is an edge in G, and zero otherwise. (Note: $\mathbf{A}$ is not the same as the ordinary $2n$ by $2n$ adjacency matrix for G.) Define a family of sets where: S equals $E(G)$ and F equals the set of perfect matchings on G. Assign weights chosen uniformly and independently from $1 .. 2|E|$ to the edges of G. Denote the weight assigned to the edge (u_i, v_j) by w_{ij}. Define an auxiliary matrix $\mathbf{B}$ by replacing each nonzero i, j-entry of $\mathbf{A}$ by $2^{(w_{ij})}$. Denote the determinant of $\mathbf{B}$ by Det($\mathbf{B}$), and the determinant of the matrix obtained from $\mathbf{B}$ by deleting its ith row and j^{th} column by Det($\mathbf{B_{ij}}$).

The following theorem gives a matrix/arithmetic characterization of the (probabilistically) unique, minimum weight, perfect matching of G.

Theorem (Characterization of Unique Perfect Matching). Let M be a unique minimum weight perfect matching in $G(U, V, E)$, and let w be its weight. Then

(1) The highest power of 2 that divides Det($\mathbf{B}$) is 2^w, and

(2) An edge (u_i, v_j) is in M if and only if

$$\text{Det}(\mathbf{B_{ij}})2^{(w_{ij})} / 2^w \text{ is odd .}$$

The proof is as follows. First, observe that every perfect matching on G induces a permutation p on $1, .., n$ given by $p(i) = j$ if and only if (u_i, v_j) is an edge of the matching. Of course, not every permutation corresponds to a perfect matching. Indeed, if we define the *Value* of a permutation p as

$$\text{Value}(p) = \prod_{i=1}^{n} \mathbf{B}(i, p(i)),$$

then Value(p) is nonzero if and only if p corresponds to a perfect matching. If M' is a perfect matching of weight,

$$\text{Weight}(M') = \sum_{i=1}^{n} w(i, p_{M'}(i)),$$

where $p_{M'}$ is the permutation determined by M', Value($p_{M'}$) equals $2^{\text{Weight}(M')}$. For the permutation p_M determined by M, Value(p_M) equals 2^w. By definition,

$$\text{Det}(\mathbf{B}) = \sum_{\text{all permutations } p} \text{sign}(p) \, \text{Value}(p).$$

By supposition, every matching other than p_M has weight strictly greater than w. Therefore, except for p_M every permutation has Value either 0 or a higher power of 2 than 2^w. Therefore, 2^w is the highest power of 2 dividing Det($\mathbf{B}$), as required by part (1) of the statement of the theorem.

To prove part (2), consider for fixed i and j the sum

$$\sum_{p \, : \, p(i) = j} \text{sign}(p) \, \text{Value}(p),$$

where the summation is taken over all permutations on $1, \dots, n$ satisfying that $p(i)$ equals j. The term $\mathbf{B}(\mathbf{i}, \mathbf{j})$ is common to every instance of Value(p) that occurs in this sum. If we factor this term out, the remaining terms correspond to those that arise from 1-1 mappings from $\{1, \dots, n\} - \{i\}$ to $\{1, \dots, n\} - \{j\}$, and so by definition equal Det($\mathbf{B_{ij}}$), whence the sum equals

$$2^{(w_{ij})} \, \text{Det}(\mathbf{B_{ij}}).$$

If (u_i, v_j) is an edge of M, then the permutation corresponding to M yields a term in the summation of value 2^w. The remaining terms are all either zero or powers higher than 2^w. Therefore,

$$\sum_{p \, : \, p(i) = j} \text{sign}(p) \, \text{Value}(p)/2^w$$

has the form

$$2^w(1 + 2k)/2^w,$$

for some integer k, which is always odd. On the other hand, if the edge (u_i, v_j) is not in M, every term in the sum is either zero or a higher power than 2^w, so the expression then has the form

$$2^{(w+1)}k/2^w,$$

for some integer k, which is always even. This completes the proof of part (2) of the theorem.

A parallel procedure based on the theorem is easily devised. The following procedure has probability at least $1/2$ of finding a perfect matching; so if repeated k times, the probability of its not finding a perfect matching is at most $(1/2)^k$. The procedure relies on fast parallel algorithms for computing the determinant and minors of a matrix. Refer to Pan (1985) for a random matrix inversion algorithm which computes these quantities for an n by n matrix with m-bit entries using $O(mn^{3.5})$ processors in $O(\log^2 n)$ time.

We assume G is represented by a bipartite adjacency matrix of type Bigraph.

```
type   Bigraph = record
                 |V(G)|: Integer
                 A(1..|V(G)|/2, 1..|V(G)|/2): Integer
               end
```

The submatrix obtained by deleting the ith row and j^{th} column of B is denoted by $\mathbf{B_{ij}}$. The type Edge is just an ordered pair.

Parallel_procedure Match (G, M)

(* Returns a perfect matching for G in M with reliability 1/2 *)

var G: Bigraph
  M: Set of Edge
  i, j: 1..|V(G)|/2
  w: Integer
  B(1..|V(G)|/2, 1..|V(G)|/2): Integer
  B_{ij}: i,j Submatrix of B
  Det: Integer function

Set M to Empty

for i, j = 1 to |V(G)|/2 **do_in_parallel**

  if A(i,j) <> 0
  then **Set** A(i,j) to a random value chosen from [1, 2|E(G)|]
     Set B(i,j) to $2^{A(i,j)}$
  else **Set** B(i,j) to 0

Set w to the highest power of 2 that divides Det(B)

for i, j = 1 to |V(G)|/2 **do_in_parallel**

  if B(i,j) Det(B_{ij}) / 2^w is odd
  then **Set** M to M ∪ { (i,j) }

End_Parallel_Procedure_ Match

The performance of the algorithm is summarized in the following theorem.

Theorem (Parallel Matching Performance). Let $G(V, E)$ be a bipartite graph containing a perfect matching. Then, the procedure Match finds a perfect matching on G in $O(\log^2 |V|)$ time on a CREW PRAM with $O(|E||V|^{3.5})$ processors.

The proof is immediate from the procedure. The determinant calculations take $O(mn^{3.5})$, where m is the length of the entries of B in bits, by the algorithm of Pan. For the present problem, the entries are at most $2^{|E|}$, so m equals $|E|$; while n equals $|V| / 2$, whence the processor estimate follows. The set union operation can obviously be implemented in an EW mannner, even though, for economy of expression, we have not stated it that way in the procedure. This completes the proof.

Of course, the parallel algorithm Match induces a random polynomial time sequential algorithm, obtained by merely performing the parallel steps by a sequential procedure.

We now consider several other related matching problems that can be solved using essentially the same technique.

Perfect matching on a graph. The preceding technique can be applied with minor changes to compute a perfect matching on an arbitrary graph $G(V, E)$. Let **T** be the Tutte-Lovasz matrix of G. We first use the Tutte-Lovasz matching condition to determine if G has a perfect matching. If it does, we randomly assign weights from $[1, 2|E(G)|]$ to the edges of G, just as in the bipartite case. Denote the weight assigned to (i, j) by w_{ij}. We then define a matrix **B**, that plays the same role as the matrix **B** in the bipartite case, by replacing each indeterminate x_{ij} in **T** by $2^{w_{ij}}$. Then, the same characterization theorem holds as in the bipartite case, though the proof is more difficult, and the same algorithm computes the matching.

Minimum weight perfect matching. We can solve the minimum weight perfect matching problem on a weighted graph $G(V, E)$, under the assumption that the weights are represented in unary, in time polynomial in the size of the problem, using $O(|V|^{3.5} |E| w_{max})$ processors, where w_{max} is the heaviest edge weight.

We first scale the given edge weights on G by multiplying each of them by $|V||E|$. This has the effect of making the weight of every minimum weight perfect matching lighter than every other perfect matching by at least $|V||E|$. We then use the isolating technique to probabilistically isolate one of the minimum weight perfect matchings: we set the weight of each edge (i, j) to w_{ij} plus a randomly chosen value from $[1, 2|E(G)|]$, just as in the bipartite case. The Isolating Lemma can be established just as before, and the perfect matching algorithm for general graphs can then be applied.

Maximum matching on a graph. We can find a maximum matching on a graph $G(V, E)$ by reducing the problem to finding a minimum weight perfect matching on a transformed graph G' defined as follows. We first add edges to G until it becomes complete. Then, we assign zero weight to the edges originally in G, and unit weight to the newly added edges. A minimum weight perfect matching on G' then determines a maximum matching on G, and conversely.

9-3 SOFTWARE PIPELINE FOR HEAPS

The shared memory model of parallel computation postulates a pool of processes which access a common memory under highly specific assumptions on the nature of the shared access (CREW, etc.), but with the coordination of the processes usually left unspecified. We now consider an example of a pool of cooperating processes, performing parallel heap creation and deletion (the heap operations used in the minimum spanning tree algorithm) which make concurrent read and write memory accesses to shared locations and that have to be carefully coordinated to obtain correct results. The parallel deletion algorithm will be coordinated using the standard P and V synchronization primitives. The parallel algorithm that creates the heap is relatively simpler, but has an interesting performance analysis. We describe the deletion algorithm first.

Parallel heap deletion. Let $\mathbf{H}$ be a heap of size n and height h. There is a standard sequential algorithm for deleting the minimum element from a heap (see Chapter 4). The parallel algorithm uses the identical compare and sift operations, but implements them using a sequence of processes which cooperate in the manner of a "bucket brigade." One process, Delete_Heap, removes the value at the root of the heap. This creates a ripple effect which is handled by other processes, Maintain_Heap, of which there is one instance per level of the heap, and which keep their delegated levels full by sifting values up from lower levels.

Each level of the heap is concurrently processed by a pair of processes: a consumer process that removes a value from the level, and a producer process that fills a node in the level when a vacancy appears. Conversely, each process concurrently processes two levels: the one it deletes from, and the one it adds to. When acting as a consumer, a process must test if a level is full before removing a value from the level; when acting as a producer, a process must wait for a vacancy before adding a value to a level.

A process must have sole access to both its levels when it is active in order to operate correctly. That is, it must not be working with data that is being concurrently accessed by another process. Therefore, before entering its critical section, a process must make complementary access tests: For its producer level, it tests that a flag indicating the level is empty is set; for its consumer level, it tests that a complementary flag indicating the level is full is set.

The synchronization flags are tested by the P and V operations, which are indivisible in the sense that no other processes can access their arguments concurrently. The flags are called semaphores. If a process invokes $P(s)$, the semaphore argument s is decremented by 1 and the issuing process is blocked if s becomes zero. If a process invokes $V(s)$, then a process blocked on s is reawakened, or if no process is blocked on s, then s is incremented by 1. We will use arrays of semaphores Full(h) and Empty(h); so there is one semaphore of each type for each level of the heap. Empty(i) is initialized to 0, and Full(i) is initialized to 1 (for $i = 1, \ldots, h$), corresponding to the assumption that every level of the heap is initially full. Empty(i) is set to 1 whenever there is an empty node at level i of the heap, and equals 0 otherwise. Full(i) is set to 1 whenever level i of the heap is full, and equals 0 otherwise.

The procedure for Delete_Heap operates as follows. Before accessing $\mathbf{H(1)}$, Delete_Heap executes P, which suspends Delete_Heap if $\mathbf{H(1)}$ is empty, in which case the level 1 maintenance routine will execute a correlative V when it fills $\mathbf{H(1)}$, which will awaken Delete_Heap if it is blocked. After accessing the value in $\mathbf{H(1)}$, Delete_Heap signals the level 1 maintenance routine that $\mathbf{H(1)}$ is empty. An array **Node** allows direct identification of vacant nodes: **Node(i)** equals the index of the empty node (if any) on level i (**Node(1)** is initialized to 1 and never reset). The heap is assumed to be a complete binary tree, filled where necessary with dummy values M. An example is shown in Figure 9-7.

Procedure Delete_Heap (H(1), Empty(1), Full(1), Min)

(* Returns H(1) in Min, after synchronization *)

var H(1), Min: Real
Full(1), Empty(1): Integer Semaphore

P(Full(1))	(* Wait for a value *)
Set Min to H(1)	(* Store it in min *)
V(Empty[1])	(* Signal its deletion *)

End_Procedure_Delete_Heap

Each maintenance procedure is responsible for keeping its own level full. Maintain_Heap (H, n, Empty, Full, Node, h, i) is the procedure for level i, and is signalled by the level $i - 1$ procedure (or Delete_Heap for level 1) whenever a node j on level i has been emptied. At that point, it waits for a signal that level $i + 1$ has been filled, fills the empty node j from level $i + 1$, signals the level $i + 1$ process that level $i + 1$ has a vacancy, and signals the level $i - 1$ process that level i is full. The operation of the level i procedure(s), $i = 1, \ldots, h-1$, is slightly different from the level h procedure, and is described first.

Procedure Maintain_Heap (H, n, Empty, Full, Node, h, i)

(* Keeps heap level i full *)

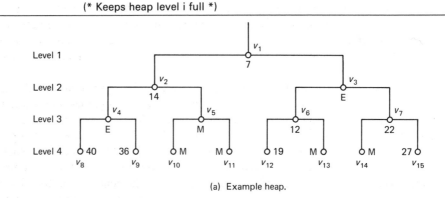

(a) Example heap.

	Full	Empty	Node
Level 1	1	0	(1)
Level 2	0	1	3
Level 3	0	1	4
Level 4	1	0	—

(b) Semaphores and node pointer.

Figure 9-7. Software pipeline heap.

```
var   n, h: Integer
      H(n): Real
      Node(h), j: 0..n
      i: 1..h
      Full(h), Empty(h): Integer Semaphore
```

repeat (forever)

```
   P(Empty(i))   (* Wait for Level i to have an empty node *)
   P(Full(i + 1)) (* Wait for Level i + 1 to be full *)

   Set j to Node(i)
   if   H(2j) < H(2j + 1) then   Set H(j) to H(2j)
                                 Set Node(i + 1) to 2j
                          else   Set H(j) to H(2j + 1)
                                 Set Node(i + 1) to 2j + 1
   Set Node(i) to 0

   V(Empty(i + 1)) (* Signal Level i + 1 has an empty node *)
   V(Full(i))         (* Signal Level i is full *)
```

End_Procedure_ Maintain_Heap(i)

The level h procedure refills the bottom level nodes with dummy values.

Procedure Maintain_Heap_H (H, n, Empty, Full, Node, h)

(* Refills bottom level with dummy values *)

```
var   n, h: Integer
      H(n): Real
      Node(h), j: 0..n
      i: 1..h
      Full(h), Empty(h): Integer Semaphore
      M: Constant
```

repeat (forever)

```
   P(Empty(h))                  (* Wait for vacancy *)
   Set Heap(Node(h)) to M
   V(Full(h))                   (* Signal full *)
```

End_Procedure_ Maintain_Heap(h)

The performance advantage of a parallel heap lies in its ability to handle deletions in a pipelined manner. This feature and the correctness of the synchronization are summarized in the following theorem.

Theorem (Synchronization Correctness and Performance).   Let **H** be a heap of height h and size n. Then, the procedures for deletion and maintenance correctly implement parallel heap deletion. Furthermore, the deletion algorithm has $O(1)$ pipeline delay.

The proof is as follows. The synchronization primitives are certainly nominally correct, since each level i procedure notifies its neighboring level $i + 1$ procedure whenever it creates a vacancy at level $i + 1$, and in turn waits for its neighboring level $i - 1$ procedure to notify it of a vacancy at level i. However, it remains to prove that access to the critical sections (the code bracketted by the P and V guards) is mutually exclusive. Thus, observe that the processes for levels i and $i - 1$ cannot access their shared data concurrently, because otherwise each of these processes would have successfully tested their synchronization semaphores: Empty($i - 1$) and Full(i) (in the case of the level $i - 1$ procedure), and Empty(i) and Full($i + 1$) (in the case of the level i procedure). In particular, the level $i - 1$ procedure would have been unblocked or passed at Full(i), while the level i procedure would have been unblocked or passed at Empty(i). But, these are contradictory conditions which cannot hold concurrently. This completes the proof of the correctness of synchronization.

Assume for simplicity that Delete_Heap and each Maintain_Heap operation take unit time. Then, if a deletion is initiated at time 0, it takes until time h before the whole heap is updated, the same as for a sequential heap. However, the difference is that subsequent deletions can be pipelined in the case of a parallel heap, that is, they can be initiated with a delay of only one unit of time, and concurrently with the ongoing updates of the higher levels of the heap. In particular, the system can perform d deletions in $O(d)$ time. This completes the proof.

Parallel heap creation. The sequential recursive procedure for constructing a heap (Chapter 4) can be directly adapted to create a heap in parallel by delegating the recursive creation of subheaps to parallel processors, instead of to a single processor as in the sequential algorithm. The performance of this algorithm is given by the following theorem.

Theorem (Parallel Heap Creation). Let S be a set of cardinality n, and let P be a set of $\lceil \frac{n}{4} \rceil$ parallel processors. Then, the parallel recursive algorithm constructs a heap on S in $O(\log^2 n)$ time.

The proof is as follows. Let $T(n, n/4)$ denote the time required to create a heap on n elements using $n/4$ processors. For simplicity, assume $n + 1$ is a power of 2. Following the sequential heap creation algorithm, we select an element x from S, and partition $S - \{x\}$ into two equal parts $S1$ and $S2$. Then, we create heaps on the elements of $S1$ and $S2$ in parallel, using $n/8$ processors for each part. The total parallel time required to create a heap on S then satisfies

$$T(n, n/4) = T(n/2, n/8) + O(\log n),$$

where $T(n/2, n/8)$ comes from the time to create heaps on Si ($i = 1, 2$) in parallel, and $O(\log n)$ comes from the time required to sequentially sift x into its correct position in the composite heap after the subheaps on $S1$ and $S2$ have been constructed. If $n + 1 = 2^k$, we can iterate the recurrence relation k times to obtain

$$T(n, n/4) = O(\log n + \log n/2 + \log n/4 + \ldots + \log n/2^k),$$

or equivalently,

$$= O(k + (k - 1) + (k - 2) + \ldots + 1),$$

which sums to $O(k(k - 1)/2)$ or $O(\log^2 n)$, which completes the proof.

9-4 TREE PROCESSOR CONNECTED COMPONENTS ALGORITHM

A *tree processor* is a system of processors interconnected by a rooted tree. Such an organization is especially suitable for divide and conquer algorithms. For simplicity, we will assume the tree is a completely balanced binary tree. Each processor can communicate with its neighbors (its parent and two children processors) except for the endpoint processors, which have no children. The processor at the root controls the other processors; otherwise, the details of interprocessor communication and synchronization are ignored. The system does all its I/O through the endpoint processors, under the control of the root processor.

Tree processors can efficiently perform the following functions. (The number of processors in the tree is denoted by n.)

(1) *Broadcasting,* that is, transmitting a value stored at the root processor to all the endpoint processors in $O(\log n)$ time.

(2) *Census Function Evaluation:* A *census function* is a commutative associative function f of values in the endpoints. Examples include: summation, logical **and**, logical **or**, and the minimum or maximum of values at the endpoints. If each binary subfunction of f is computable in time t, then f itself can be computed at the root in time $O(t \log n)$.

(3) *Selection,* that is, retrieving to the root processor the value of a field in an endpoint processor in $O(\log n)$ time.

We will illustrate the use of a tree processor of order $2|V| - 1$ to compute the components of a graph $G(V, E)$ from its adjacency matrix $\mathbf{A}$, using a successive component merging technique. $\mathbf{A}$ is read a row at a time into the endpoints of the tree processor, each row causing more estimated components to merge. The jth endpoint processor handles vertex j, and contains registers: *Component,* which stores the current estimated component number of the vertex; *Index,* which gives the number of the vertex; and *Select,* which stores the current adjacency matrix entry for the vertex. We follow the convention that the number of a component is equal to the least index of any vertex in the component. After $\mathbf{A(i, j)}$ is read into Select(j) at endpoint j at iteration i, the algorithm finds the smallest component number c of any vertex whose Select bit is set, and then successively identifies the component numbers r of each other selected vertex, and merges all vertices with component numbers matching r, whether selected or not, into the c component. On termination, Component(j) gives the number of the component that vertex j belongs to.

Figure 9-8 illustrates the merging technique. It shows partial components, numbered c, $c1$, and $c2$ (c is the lowest selected component number), recognized prior to reading the ith row of the adjacency matrix. The components are merged in Figure 9-8b by setting to c the component numbers of all vertices in every partial component adjacent to the c component.

An example is given in Figures 9-9 and 9-10. First, a census operation brings the number of the smallest selected component c ($= 1$) to the root processor, whence it is broadcast to all the endpoints. The components of selected vertices are then merged one at a time into component c by retrieving the component number r of some selected vertex to the root and then making each endpoint processor with a vertex lying in component r, whether selected or not, merge with c by changing its component number to c

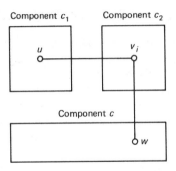

(a) Components prior to reading row i.

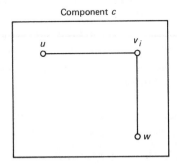

(b) Component after reading row i.   **Figure 9-8.** Component merging.

and turning off its Select field. The process is repeated for each distinct component adjacent to vertex v_1 as shown in Figure 9-10b. Figure 9-10b also shows the next row (row 2) of the adjacency matrix as well as the next value of $c(= 2)$ at the root. The process continues until Figure 9-10e, when all the components have been merged. The last row of **A** leads to no further changes.

The procedure Find_Components implements the algorithm. The boolean function Get(Selected_component) returns, in Selected_component, the component number of a vertex whose Select field is on, and fails if there is none. The graph $G(V,E)$ is represented by its adjacency matrix.

Parallel Procedure Find_Components (G, Component)

(* Uses tree processor with $2|V| - 1$ processors to find the
    components of G *)

var G: Graph
    Component(1..|V|), Selected_component, i, c: 1..|V|
    Select(1..|V|): 0..1
    Get: Boolean function

Broadcast: Set Component(i) to i, at each endpoint processor i

v_1 v_5 v_4 v_2 v_3 **Figure 9-9.** $G(V,E)$ for tree processor example.

○———○———○———○———○

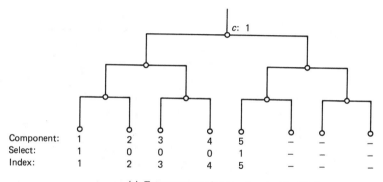

Component:	1	2	3	4	5	—	—	—
Select:	1	0	0	0	1	—	—	—
Index:	1	2	3	4	5	—	—	—

(a) Tree processor after row 1 and *c* identified.

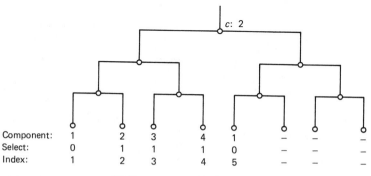

Component:	1	2	3	4	1	—	—	—
Select:	0	1	1	1	0	—	—	—
Index:	1	2	3	4	5	—	—	—

(b) Tree processor after row 2 and next *c* identified.

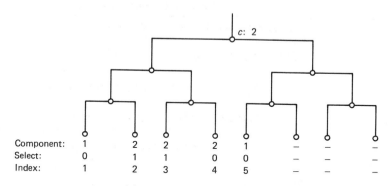

Component:	1	2	2	2	1	—	—	—
Select:	0	1	1	0	0	—	—	—
Index:	1	2	3	4	5	—	—	—

(c) Tree processor after row 3 and next *c* identified.

Figure 9-10. Trace of **Find_Components** for $G(V, E)$.

repeat |V| times

Broadcast: To each endpoint i: Read next adjacency matrix
value into Select(i)

Census: Get smallest component number c
among all endpoints i with Select(i) = 1

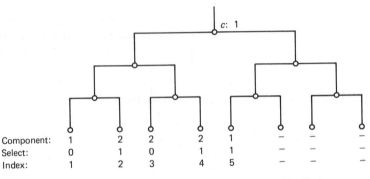

Component:	1	2	2	2	1	–	–	–
Select:	0	1	0	1	1	–	–	–
Index:	1	2	3	4	5	–	–	–

(d) Tree processor after row 4 and next c identified.

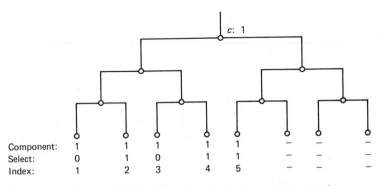

Component:	1	1	1	1	1	–	–	–
Select:	0	1	0	1	1	–	–	–
Index:	1	2	3	4	5	–	–	–

(e) After effect of row 4 with all components merged.

(Figure 9-10 continued)

Broadcast: Transmit c to all endpoints

while Get (Selected_component) **do**

Broadcast: Send Selected_component to all endpoints

Broadcast: At each endpoint processor i
if   Component(i)   = Selected_component
then **Set** Component(i) to c
Set Select(i) to 0

End_Procedure_Find_Components

The results are summarized in:

Theorem (Tree Processor Components).   Let $G(V, E)$ be a graph, and let T be a complete binary tree processor of order $2|V| - 1$. Then, Find_Components computes the components of G in $O(|V| \log |V|)$ time.

The proof of correctness is straightforward. The performance of the algorithm is determined by the number of broadcast, census, and selection operations done, which in turn depend on the number of components merged. There are at most $|V|$ components; so there are at most $|V|$ nontrivial merging actions. Each merge requires $O(\log |V|)$ steps;

so the overall algorithm takes $O(|V| \log |V|)$ time. The performance is dependent on whether the tree processor can initiate reading the next row of the adjacency matrix, or must wait for the arrival of the rows at predetermined intervals. This completes the proof.

9-5 HYPERCUBE MATRIX MULTIPLICATION AND SHORTEST PATHS

Hypercubes were introduced in Chapter 6, where techniques for routing data on them were considered. The hypercube architecture presented here has the same underlying topology as the earlier model, but is more restrictive in the way it allows data to be routed, and has a centralized system of control. We show how to multiply matrices on this model. The algorithm is suitable for iteration and so can also be used to efficiently compute powers of matrices, allowing it to be applied to graph-theoretic problems like shortest paths and transitive closure (see the exercises).

A hypercube, like any parallel computer, is determined by a few basic features: its control mechanism, the functions of its individual processors, its interconnection network, and the ways in which it can route data through its network. Topologically, an *m-dimensional hypercube* (or *m-cube*) is a collection of 2^m processors interconnected in the same manner as the vertices of an m-dimensional cube. The processors each have m bit addresses, a processor with address $(p_{m-1}, p_{m-2}, \ldots, p_0)$ being directly connected to each of the m processors whose addresses differ from its address in a single bit. Figure 9-11 shows a three-dimensional hypercube. Figure 9-12 shows part of a four-cube, with just one of its eight fourth-dimensional connections shown.

Each vertex processor executes the same program on its own local data, but under the control of the single external control processor, an arrangement which is called a *Single Instruction Multiple Data Stream* or SIMD computer. The control processor can selectively activate processors. For example, it can select a set of processors whose addresses satisfy an address mask constraint and command them to execute the next step of the common program, or route data to their neighbors.

There are *m data routing functions* denoted R_i, $i = 0, \ldots, m-1$. R_i transmits data between the processors p and p^c with addresses which are complementary in the bit i:

$$R_i: (p_{m-1}, \ldots, p_i, \ldots, p_0) \longleftrightarrow (p_{m-1}, \ldots, p_i^c, \ldots, p_0),$$

where p_i^c denotes the complement of p_i. Geometrically, R_i moves data between processors adjacent along the ith dimension of the hypercube. As many as 2^{m-1} pairs of proces-

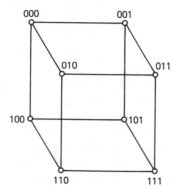

Figure 9-11. Three-cube on eight vertices.

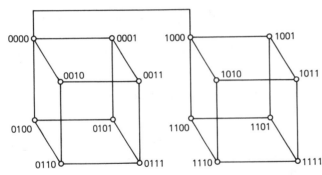

Figure 9-12. Four-cube with only one fourth-dimensional connection shown.

sors can transmit data simultaneously, but because the routes define a matching on the graph of the hypercube, there is no possibility of access contention for an edge. Only processors currently selected by the control processor participate in the data movement. Despite the limited nature of the interconnections, no processor is more than m data movements away from any other processor, since the processor addresses are only m bits long.

Matrix multiplication. Since the procedure for computing the product of a pair of matrices **A** and **B** is straightforward, the only issue is how to do so in a manner that capitalizes on the parallel computing ability of the hypercube. This necessarily entails distributing the problem data around the hypercube to take advantage of its parallelism. Our strategy will be based on the observation that the components of the product **AB** are the dot products of the rows of **A** and the columns of **B**. There are N rows of **A** and N columns of **B**, and so there are N^2 dot products to calculate. Given a hypercube with N^2 processors, we could in principle, calculate all N^2 dot products in parallel, provided the data and computations could be distributed properly. The problem is simplest if we assume $N = 2^m$, in which case the $2m$-dimensional hypercube has exactly N^2 processors, precisely the number needed for optimal parallel computation.

The hypercube matrix multiplication algorithm has several phases.

(1) Initialization
(2) Distribution
(3) Computation
(4) Inverse Distribution

We will skirt the details of initialization. Presumably, the control processor can load **A** and **B** to a set of vertex processors. This may be a bottleneck if the data is loaded sequentially. However, as we shall see, if the multiplication is part of a sequence of multiplications (such as, for computing $\mathbf{AB}^k$, for some positive integer k), then this setup cost may be amortized over the whole sequence. We do make the assumption that the rows of **A** and columns of **B** are initially distributed in a manner, which we will specify later, which facilitates the distribution requirements of the next phase.

Distribution refers to the dissemination of the matrix data around the hypercube so the data can be processed in parallel. Denote the ith row of **A** by $\mathbf{A_i}$, and the jth column of **B** by $\mathbf{B_j}$. Each row of **A** and column of **B** is used in N dot products; so we need to broadcast multiple copies of each of them around the hypercube, using an appropriate sequence of hypercube routing functions, in such a way that copies of both row A_i and column B_j end up at the processor whose address is:

$$(j - 1)|(i - 1),$$

where "|" denotes concatenation and $(j - 1)$ and $(i - 1)$ are m bit address strings.

Once the data is in position, the computation of the dot products can be done in parallel at the vertex processors in $O(N)$ time. Note that the unit of time here is with respect to the computation rate of the processors; while the time for data distribution is measured in terms of the transmission speed, which may typically be an order of magnitude (at least) slower than the computation rate.

The final phase is inverse distribution. Once the computation phase is done, we inversely broadcast the individual components of the matrix product back to the home locations of the rows of A, using the inverse of the broadcasting sequence used to distribute the copies of A around the hypercube in step (2). (Distribution is commonly called *scattering*, and inverse distribution is called *gathering*.) The inverse distribution gathers the rows of the product back at the home locations of the corresponding rows of A, allowing us to iterate the multiplication algorithm without reloading the data. In the special case where A and B are equal, the algorithm can be applied iteratively to compute powers of A, as required for example to compute the transitive closure of a digraph.

We will now examine each phase in detail.

Initialization. We assume the matrices are initially loaded so the ith row of A is located at the processor with the address

$$(0)|(i - 1).$$

If the binary representation of $i - 1$ is $(i_{m-1}, \ldots, i_0)$, this is equivalent to

$$(i_{2m-1} = 0, \ldots, i_m = 0 | i_{m-1}, \ldots, i_0).$$

We assume the jth column of B is initially located at

$$(j - 1)|(0).$$

If the binary representation of $j - 1$ is $(j_{m-1}, \ldots, j_0)$, this is equivalent to

$$(j_{m-1}, \ldots, j_0 | 0, \ldots, 0).$$

Refer to Figures 9-13 and 9-16 for an example where $m = 2$ and $N = 4$. The 4 $(= 2^m)$ rows of A are initially located at the vertices starred in Figure 9-13. The columns of B are initially at the vertices starred in Figure 9-16.

Distribution. The matrices are distributed using a tree-like broadcasting algorithm. The routing functions used to implement the distribution are

$$R_m \ldots R_{2m-2} R_{2m-1},$$

which broadcast the initial copy of A_i located at $(0)|(i - 1)$. The routing functions used to implement the B broadcast are

$$R_0 R_1 \ldots R_{m-1},$$

which broadcast the initial copy of B_j located at $(j - 1)|(0)$.

The procedure for broadcasting A is illustrated in Figures 9-13, 9-14, 9-15 and for B in Figures 9-16, 9-17, 9-18. In Figure 9-14, m equals 2 and the broadcasts for A

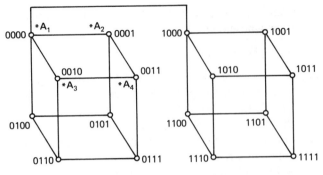

Figure 9-13. Initial distribution of rows of **A**.

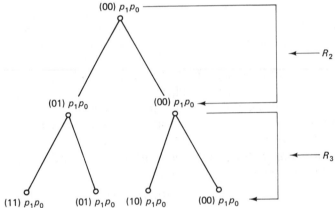

Figure 9-14. Broadcast of **A** row with home address at $00p_1p_0$.

are R_2 and R_3. In Figure 9-17, the broadcasts for **B** are R_0 and R_1. Figures 9-14 and 9-17 represent the broadcast as a binary tree propagation scheme. Observe that despite the increasing number of vertices that become involved as the propagation scheme progresses, there is never any chance of congestion.

Row A_i, where $i - 1 = (i_{m-1}, \ldots, i_0)$ in binary, is broadcast to all processors with addresses of the form

$$(x_{m-1}, \ldots x_0, i_{m-1}, \ldots, i_0),$$

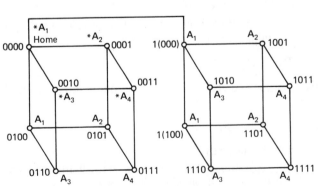

Figure 9-15. Distribution of rows of **A** after broadcast.

Parallel Algorithms   Chap. 9

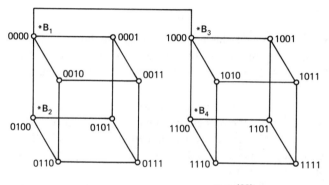

Figure 9-16. Initial distribution of columns of **B**.

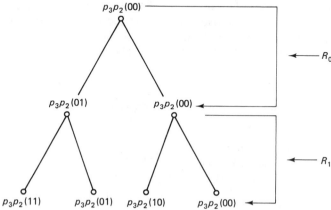

Figure 9-17. Broadcast of **B** column with home address at $p_3p_2(00)$.

where the indices x_i run over all possible combinations of 0s and 1s; so each row is distributed to $2^m (= N)$ locations. On the other hand, column $\mathbf{B}_j$, where $j - 1 = (j_{m-1}, \ldots, j_0)$ in binary, is broadcast to all processors with addresses of the form

$$(j_{m-1}, \ldots, j_0, x_{m-1}, \ldots, x_0).$$

The actual order of the broadcasts is unimportant because it only affects the order in which the processors receive the data, not the set of processors that receive the data. Upon completion of the broadcasting phase, the pair $\mathbf{A}_i$ and $\mathbf{B}_j$ needed to compute $\mathbf{C}_{ij}$ are coresident at the processor with address,

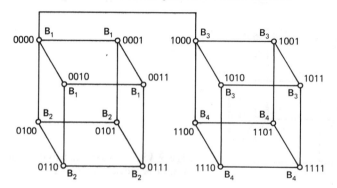

Figure 9-18. Distribution of columns of **B** after broadcast.

$$(j_{m-1}, \ldots, j_0, i_{m-1}, \ldots, i_0) .$$

The m leading bits of this address match the leading m bits of the home address of $\mathbf{B_j}$, while its trailing m bits match the trailing m bits of the home address of $\mathbf{A_i}$. Refer to Figure 9-19 for an example where $m = 2$. The $\mathbf{A_i B_j}$ dot product is then computed at the host processor.

Inverse Distribution. The N processors at which the dot products of $\mathbf{A_i}$ and $\mathbf{B_j}$ (i fixed, $j = 1 .. N$) are computed are exactly the processors to which the initial copy of $\mathbf{A_i}$ was broadcast originally. Therefore, applying the inverse sequence of broadcasts R_{2m-1}, $R_{2m-2}, \ldots, R_m$ gathers the distributed components of the ith row $\mathbf{C_i}$ of the matrix product back to the home processor of $\mathbf{A_i}$. Referring to Figure 9-19, for example, the components of the first row of $\mathbf{C}$ lie at the starred processors, which are the processors to which $\mathbf{A_1}$ was originally broadcast from its home location (0000). The inverse broadcast sequence merely gathers them back to (0000).

The procedure Matrix_Product_on_Hypercube for computing AB on a hypercube follows.

Procedure Matrix_Product_on_Hypercube (A,B,C)

(* Returns in C the product of N by N matrices A and B, using
   the 2m-cube (where $2^m = N$) *)

var   N: Constant
         A(N,N), B(N,N), C(N,N): Real
         m: Constant = lg N
         i, j: 1..N
         k: 0..2m − 1
         A(1..N): Row of A
         B(1..N): Column of B
         p(0..2m − 1): 0..1

(* Initial locations:

    A_i, where $i - 1 = (p_{m-1}...p_0)$ in binary, lies at:

        $(p_{2m-1} = 0, \ldots, p_m = 0 \mid p_{m-1}, \ldots, p_0)$

    B_j, where $j - 1 = (p_{2m-1}...p_m)$ in binary, lies at:

        $(p_{2m-1}, \ldots, p_m \mid p_{m-1} = 0, \ldots, p_0 = 0)$ *)

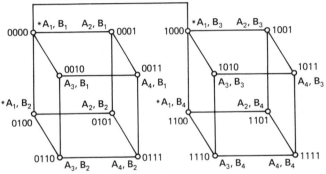

Figure 9-19. Final distribution of rows of **A** and columns of **B**.

(* Broadcast rows of A *)

for k = m to 2m − 1 **do**

    for each processor p with address bits: $p_k = 0, \ldots, p_{2m-1} = 0$

      do_in_parallel

      (* Send rows of A at selected processors p to $R_k(p)$ *)

      Route A row at:

$$(p_{2m-1} = 0, \ldots, p_{k+1} = 0, |p_k = 0|, p_{k-1}, \ldots, p_m, p_{m-1}, \ldots, p_0)$$

      to

$$(p_{2m-1} = 0, \ldots, p_{k+1} = 0, |p_k = 1|, p_{k-1}, \ldots, p_m, p_{m-1}, \ldots, p_0)$$

(* Broadcast columns of B *)

for k = 0 to m − 1

    for every processor p with $p_k = 0, \ldots, p_{m-1} = 0$

    do_in_parallel

      (* Send columns of B at selected processors p to $R_k(p)$ *)

      Route B column at:

$$(p_{2m-1}, \ldots, p_m, p_{m-1} = 0, \ldots, p_{k+1} = 0, |p_k = 0|, p_{k-1}, p_0)$$

      to:

$$(p_{2m-1}, \ldots, p_m, p_{m-1} = 0, \ldots, p_{k+1} = 0, |p_k = 1|, p_{k-1}, p_0)$$

(* Compute row / column dot product at each processor *)

for each processor with address (j − 1) | (i − 1) **do_in_parallel**

$$\text{Set} \quad C(i,j) \text{ to } \sum_{k=1}^{N} A(i,k)B(k,j)$$

(* Gather C rows to home processors for A by inverse broadcast *)

for k = 2m − 1 to m **do**

    for every processor p with $p_r = 0$, for r > k and $p_k = 1$

      do_in_parallel (* Send components of C at p to $R_k(p)$ *)

      Route C component at:

$(p_r = 0 \text{ for } r > k, p_k = 1, p_{k-1}, \ldots, p_m, p_{m-1}, \ldots, p_0)$ to:

$(p_r = 0 \text{ for } r > k, p_k = 0, p_{k-1}, \ldots, p_m, p_{m-1}, \ldots, p_0)$

End_Procedure_Matrix_Product_on_Hypercube

The performance of the algorithm is summarized in the following theorem.

Theorem (Hypercube Multiplication Performance). Let H be an m-dimensional hypercube, and let **A** and **B** be a pair of N by N matrices (where $N = 2^m$). Then, the matrix product of **A** and **B** can be computed on H in $O(t_{\text{trans}} N \lg N + t_{\text{comp}} N)$ time, where t_{trans} is proportional to the time to transmit a fixed number of bits between a pair of adjacent processors on H, and t_{comp} is proportional to the instruction execution time for a vertex processor.

The proof is straightforward. There are $O(\lg N)$ broadcast and inverse broadcast steps. The broadcast steps each transmit N matrix components between pairs of adjacent processors. The inverse broadcast steps initially transmit single component values, but by the last step have aggregated these components into complete rows of the product matrix; thus $O(N)$ bits are transmitted on average per adjacent processor pair per step. Assuming the dot products are computed by sequential processors at the vertices of H, their computation time is proportional to $O(N)$. This completes the proof.

REFERENCES AND FURTHER READING

For a general survey of parallel graphs algorithm, see Quinn and Deo (1984). For an introduction to parallel computation, see Kung (1980). Lakshmivarahan (1984) gives an excellent introduction to parallel sorting algorithms and sorting networks. Akl (1985) examines parallel sorting algorithms on a variety of parallel architectures: hypercubes, perfect shuffle networks, etc. Hwang and Briggs (1984) is a very comprehensive introduction to parallel architectures, with many interesting examples.

See Kung (1982) for systolic architectures, and Molodovan (1983) for a systematic methodology for converting algorithms into systolic systems. Ullman (1984) describes a more sophisticated shortest path systolic array algorithm. An application of Molodovan's method to the design of a shortest path systolic array algorithm is given in Bhabani, et al. (1986). See also Lang, et al. (1985). Mead and Conway (1980) has many matrix processing examples on different types of systolic arrays, and is the basis of several of the exercises. Stone (1971) is a classic exposition of another special parallel architecture, the perfect shuffle.

The SM algorithms for shortest paths are from Paige and Kruskal (1985). For efficient simulation of shared memory types of access, see Vishkin (1984). The connected components algorithm is due to Shiloach and Vishkin (1982). The hypercube matrix multiplication discussion is partly based on Hwang and Briggs (1984). The parallel random matching algorithm is from Mulmuley, Vazirani, and Vazirani (1987). The random parallel matrix inversion algorithm used by them is given in Pan (1985). See Bentley (1980) for the parallel minimum spanning tree algorithm referred to in the exercises, and Luby (1986) for the suggested maximal independent set random parallel algorithm. Both Luby (1986) and Karp and Wigderson (1984) show how to transform randomized algorithms into deterministic ones.

EXERCISES

1. Design a systolic algorithm on a linear systolic array to multiply a pair of polynomials, $a_0 + a_1x + a_2x^2 + \ldots + a_nx^n$ and $b_0 + b_1x + b_2x^2 + \ldots + b_mx^m$. Analyze the performance of the algorithm.

2. Consider the solution of the linear system of equalities

$$\mathbf{A}x = b,$$

where $\mathbf{A}$ is an n by n nonsingular band lower triangular matrix of bandwidth q, that is, the first q components in column 1 can be nonzero, the second q components in column 2 can be nonzero, and so on. The solution vector $\mathbf{x}$ can be found using the recurrence relations

$$y_i^{(1)} = 0,$$

$$y_i^{(k+1)} = y_i^{(k)} + a_{ik}x_k,$$

$$x_i = (b_i - y_i^{(i)})/a_{ii}$$

We can compute the successive y's using the systolic array shown in Figure 9-20, for the case of bandwidth $q = 4$. The y_i's are input as 0 (corresponding to $y_i^{(1)}$), which are fed into the system from the rightmost processor, whence they are accumulated as they proceed through the system and then used to compute the x_i's. The x_i's emerge from the leftmost processor PE_0. The $\mathbf{A}$ coefficients enter from the top in a staggered fashion, while the b_i's enter PE_0 from the bottom. The PE_is for $i \geq 1$ are accumulators of the same kind as were used for the matrix multiplier array; however, the PE_0 is special and computes $(b_i - y_i)/a_{ii}$, which is then output as x_i. After carefully identifying the correct timing of the inputs, prove the correctness of the system, and determine the time when the final result is output.

Figure 9-20. Systolic array for triangular linear system.

3. Write a simulation that traces the operation of the systolic matrix multiplier, and keeps track of such statistics as the number of processors doing useful computations at each point in time and the completion times of the various components of the product.

4. Try to design a SM version of Ford's shortest paths algorithm.

5. Try to design a randomized SM algorithm for finding a k-coloring of a graph, where there is a processor at each vertex of the graph that can set the color of its own vertex and read access the colors of its neighboring (vertex) processors. Allow randomness in the selection of colors, and compare the performance of your algorithm with that of a backtracking algorithm for the same problem.

6. The Characterization Theorem for perfect matchings breaks down when the minimum weight matching is not unique. What happens in that case?

7. Establish the version of the Isolating lemma used in the minimum weight perfect matching problem.

8. How many edges does a k-dimensional hypercube have? What are its vertex and edge connectivity, its diameter, girth, vertex and edge chromatic number?

9. Adapt Floyd's shortest path algorithm for a 2^{n+k} order graph to a k-dimensional hypercube. Analyze the communication and computation costs.

10. Implement the explicit coordination for the heap creation algorithm.

11. "Implement" the parallel random matching algorithm, but use a sequential algorithm (like Gaussian elimination) to calculate the required determinants.

12. Write an algorithm that finds a maximum set of vertex disjoint paths between a given pair of sets, using the random perfect matching algorithm, and the problem transformation described in Chapter 2.

13. The hypercube matrix multiplication algorithm computes terms of the form:

$$\sum_{k=1}^{N} a_{ik} b_{kj},$$

in parallel. Floyd's shortest path algorithm computes terms of the form min $\{a_{ij}, a_{ik} + a_{kj}\}$. Show how to use the data distribution technique of the hypercube matrix multiplication algorithm to implement Floyd's algorithm. Do the same thing for transitive closure.

14. Write an emulation of the parallel connected components algorithm, taking care to simulate concurrent reads and writes properly. Compare the behavior of the algorithm as a function of the random resolution of concurrent writes.

15. Design a tree processor for the minimum spanning tree algorithm. Use the Prim version of the MST algorithm (used for the Voronoi MST in Chapter 4). Assume the tree processor has $|V|/(\log|V|)$ endpoint processors, each of which handles $\log |V|$ vertices in parallel or sequential fashion (in the same spirit as the parallel Dijkstra's algorithm). At each iteration, a new vertex, not yet in the minimum spanning tree is fed to the system. Each of the endpoint processors then determines for its own subset of vertices, the closest MST vertex u to the new vertex v, and the weight of (u, v). These estimates are then passed up the tree, and the closest MST vertex to v is identified at the root processor. It is then added to the tree, and the cycle is repeated with the next nontree vertex. Carefully specify the operation of this algorithm, and its performance (Bentley (1980)).

16. Consider the problem

   Problem: Maximal Independent Set
   Input: Graph $G(V, E)$
   Output: A maximal independent set I in $V(G)$

The following random parallel algorithm computes I.

```
Set I to the empty set
Set G' to G(V,E)
while G' not Empty do Get_Ind(G',I')
                     Set I to I ∪ I'
                     Set X to I' ∪ Nghb(I')
                     Set G' to the induced subgraph on V(G') − X
```

Nghb(I') is the set of vertices i in V' such that there is some j in I' for which (i,j) is in $E(G')$. The procedure Get_Ind(G',I') returns a vertex independent set of G' in I'. A parallel Monte Carlo implementation of Get_Ind is defined as follows. Let $d(i)$ equal the degree of i restricted to G'. For each i in V', let coin(i) be a 0-1 random function that takes on the value 1 with probability $1/2d(i)$ if $d(i) \geq 1$, and equals 1 always if $d(i)$ is 0. Then, the algorithm for Get_Ind(G',I') is

```
Set Z to empty
Compute in parallel for all i in V': d(i)
Compute in parallel for all i in V':
        if coin(i) = 1  then Set Z to Z U {i}
Compute in parallel for all (i,j) in E(G'):
        if i and j in Z
        then   if d(i) ≤ d(j)
               then   Set I' to I' − {i}
               else   Set I' to I' − {j}
```

Get_Ind can be implemented on an EREW PRAM using $|E|$ processors in $O(\log |V|)$ time. The while loop terminates in expected time $O(\log |V|)$, for a total expected time of $O(\log^2 |V|)$ (see Luby (1986)). Prove that I is a maximal independent set in G when the algorithm terminates.

10

Computational Complexity

The complexity of a combinatorial problem is the time it takes an algorithm to solve the problem. Sometimes, a conceptually difficult problem, like determining whether a graph is planar, has an efficient algorithm, in the case of planarity an algorithm that runs in linear time. Sometimes a conceptually simple problem, like determining the minimum number of colors needed to color a graph is (or to date appears to be) impossible to solve efficiently, since all known algorithms for the problem require an exponential amount of time. This chapter provides a framework for the consideration of the issues of complexity by introducing a precise notion of algorithm performance, and the major complexity dichotomy for combinatorial problems, the class of problems P, that can be solved in polynomial time, and the class of NP-Complete problems, that appear to be intrinsically unsolvable by any polynomial time algorithm. We also discuss pseudopolynomial, random, and parallel algorithms.

10-1 POLYNOMIAL AND PSEUDOPOLYNOMIAL PROBLEMS

It is customary in the theory of complexity to measure the performance of an algorithm for a problem as a function of the size of the input to the problem, where the input is assumed to be represented in some reasonable manner, such as by a binary representation. An algorithm which takes an amount of time which is a polynomial function of the size of its input is said to have *polynomial (time) performance;* while, a problem which can be solved by an algorithm whose performance is polynomial is said to belong to the *class of polynomial problems P*.

This classification has the convenient algebraic property of closure. For example, consider an algorithm A_1 which makes a linear number of calls to a linear time algorithm A_2. Then, A_1 is quadratic, not linear. Thus, the class of linear time algorithms is not closed under composition of algorithms. However, if A_1 makes a polynomial number of calls to a polynomial time algorithm A_2, then A_1 is also polynomial time. Thus, the class of polynomial time algorithms is closed under composition of algorithms. Practically speaking, of course, algorithms with a low polynomial order of performance (linear, quadratic, or cubic) are preferred.

The following examples illustrate this terminology.

Problem: Integer Addition
Input: A pair of integers M and N
Output: The sum $M + N$

The inputs to addition are the binary representations of the summands. The size of the input is $O(\log N)$, assuming without loss of generality that N is the larger of the two inputs. We can perform addition by sequentially adding pairwise bits of the representations of M and N with carries; so addition is solvable in $O(\log N)$ time. This is linear in the size of the input, so integer addition is in P.

Integer multiplication is also in P.

Problem: Integer Multiplication
Input: A pair of integers M and N
Output: The product MN

The length of the input for this problem is also $O(\log N)$. The standard algorithm for multiplication requires $O(\log^2 N)$ operations. As it happens, there are even faster algorithms for integer multiplication, but since $\log^2 N$ is already polynomial (quadratic) in the length of the input, we can certainly classify integer multiplication as in P.

We have previously classified algorithms such as Dijkstra's shortest path algorithm as polynomial, and we now confirm that this usage conforms to our current definition.

Problem: Vertex to Vertex Shortest Path
Input: A digraph G with non-negative, real-valued edge weights and a pair of vertices u and v in G
Output: The length of a shortest path from u to v

If suitable data structures are used, Dijkstra's algorithm solves this problem in $O(|E| \log |V|)$ steps. However, each step requires arithmetic operations on the edge weights, whose representations have to be considered part of the problem input. If the representations of the weights are uniformly bounded in length, then the required arithmetic operations will take constant time, so the existing bound still holds. Even if we allow arbitrary edge weights, the arithmetic operations take an amount of time which is linear in the size of the input, and so the algorithm is still polynomial in the length of the input: $O(|E|(\log|V|)(\log(M)))$, where M is the value of the largest edge weight. Thus, our earlier polynomial classification of Dijkstra's problem is consistent with our current usage.

It is not always easy to determine whether a problem is polynomial. The problem of maximum matching is a case in point. Its status was uncertain until it was shown to be polynomial by the algorithm of Edmonds. An example of a problem whose status is still unresolved is

Problem: Primality
Input: A positive integer N
Output: Yes, if N is prime, and no, otherwise

While Primality is conjectured to be polynomial, its computational status is still unknown. We can easily test if there is a divisor of N less than N in $O(N \log^2 N)$ steps, since each division takes at most $O(\log^2 N)$ steps. This approach seems to be polynomial in N. But, remember that performance must be polynomial in the *size* of the input to the problem, here the length of the representation for N, not the *value N* of the input. Since the performance is not polynomial in $\log N$, this algorithm is not polynomial. This does not preclude primality being polynomial, but no polynomial algorithm for primality is known. However, there is an $O((\log N)^{\log(\log N)})$ algorithm for Primality, which is almost a polynomial order of complexity. There is also an algorithm which is polynomial, provided the Extended Riemann hypothesis is true (see Cook (1983) for references). Finally, there is a simple polynomial time random algorithm for primality, which we will describe later.

Primality is an example of a *pseudopolynomial* problem, that is, a problem solvable by an algorithm whose performance is polynomial in the value of the input(s) (as opposed to being polynomial in the lengths of the representations of the inputs). While it might seem that having a pseudopolynomial algorithm for a problem is almost as good as having a polynomial algorithm for a problem, the difference is vast. For example, Dijkstra's polynomial shortest path algorithm can easily handle graphs with hundreds of vertices. But, a pseudopolynomial algorithm for primality is useless for testing the primality of numbers with several hundred digits, for although the input lengths are relatively short, the value of the input is enormous. The primality of such large numbers has practical application in cryptography.

Linear programming is another problem whose computational status was long open, but which has recently been shown to be polynomial.

Problem: Linear Programming

Input: An integer matrix $\mathbf{A}$, and an integer vector $\mathbf{b}$

Output: Yes, if there is a real vector $\mathbf{x}$ with $\mathbf{Ax} <= \mathbf{b}$, else no.

The simplex method is the traditional algorithm for solving linear programs. This algorithm has exhibited very good performance in practice, though its theoretical, worst-case performance is known to be nonpolynomial. Recently, methods with polynomial theoretical bounds have been developed, though their practical performance is another matter. The ellipsoid method of Khachian was the first polynomial algorithm for linear programming. Khachian's algorithm constructs a converging sequence of ellipsoids, each containing the optimal solution, and has performance $O(mn^3L)$, where m is the number of inequalities, n the number of variables, and L the number of bits in the input. Although it is theoretically faster than the simplex algorithm, Khachian's algorithm is much slower in practice. Karmarkar has developed a significant practical improvement based on different techniques. The Karmarkar algorithm has polynomial performance better than $O(L(mn)^2)$.

Deterministic Turing machines. A Turing Machine is a simple conceptual model of a computer that provides a rigorous framework for the theory of the complexity of algorithms. There are other models, such as Random Access Machines (RAMs) (see Aho, Hopcroft, and Ullman [1974]), but they are polynomially equivalent to the Turing model, in the sense that any computation on one model can be simulated on the other model at a polynomial cost. We shall now give a formal definition of the class of polynomial problems P in terms of Deterministic Turing Machines of polynomial time performance.

A *Turing Machine* (TM) consists of a *control unit,* that guides the operation of the machine, and a *storage unit,* which is a semi-infinite *read/write tape.* The tape consists of a sequence of cells each of which can store a single character of a *finite alphabet A.* The control unit can be in any one of a finite number of *states S.* It can read or write the tape a single cell at a time, and can move one cell to the left or right on the tape from its current position or remain stationary. The control unit uses a *Turing Table* to determine how to change its states, what to write on the tape, and how to move, in response to the current state of the unit and the character currently being read. The table consists of mappings from pairs in $S \times A$ to triples in $S \times A \times \{0, +1, -1\}$, the last corresponding to the possible movements of the control unit. The Turing Machine starts by reading the input into the first (n) cells of the tape and returning to the initial cell and an initial state, at which point it begins to execute its program on the input. It continues to execute until it reaches a (state, character) pair for which no action is specified in the Turing Table, at which point it *halts.* The output from the program is the nonempty part of the tape. If the final state of the TM is a special *accepting* state, the input string is said to be accepted by the TM. The set of all possible accepted inputs is called the language L accepted or *recognized by the TM.*

If the mappings of the Turing Table are single-valued, the TM is called a *Deterministic Turing Machine* (DTM). (If the mappings are not single-valued (that is, some pair in $S \times A$ maps to a set of triples in $S \times A \times \{0, +1, -1\}$ rather than just a single triple), the TM is called a Nondeterministic Turing Machine. We shall consider Nondeterministic Turing Machines further in the next section, when we define the class of polynomial time verifiable problems.) The *space complexity* of a TM is defined as the maximum distance, in number of cells, that the contol unit travels from the initial cell. The *time complexity* T(n) of a Deterministic Turing Machine M, is defined as a function of the length n of the inputs to the TM, and equals the maximum number of steps the TM takes over all inputs of length n. For example, if for some input of length n the TM never halts, then T(n) is infinite for that value of n.

The class of *polynomial problems* P is defined as the collection of all languages L on a finite alphabet A for which there is a Deterministic Turing Machine M such that the time complexity $T_M(n)$ of M is bounded by a polynomial $p(n)$. It is convenient, customary, and, as we shall see later, entails no loss of generality, to restrict ourselves to *decision problems,* that is, problems that have yes/no answers. Thus, we say a decision problem Q is in P if the set of instances of Q which have yes answers is precisely the language accepted by some DTM that has polynomial time complexity.

10-2 NONDETERMINISTIC POLYNOMIAL ALGORITHMS

Many combinatorial and graph-theoretic problems have the property that they are hard to solve, but easy to verify. When verification can be done in a polynomial number of steps, we say a problem is *polynomial time verifiable.* **NP** refers to the class of decision problems for which affirmative answers are polynomial time verifiable. **Co-NP** refers to the class of decision problems for which negative answers are polynomial time verifiable. It is assumed that the answers being verified are accompanied by supporting information or evidence which is available to the verification procedure. We shall now give some examples of this terminology, and then introduce a more formal definition of NP in terms of Nondeterministic Turing Machines.

Hamiltonicity is a classic instance of an NP problem:

Problem: Hamiltonicity
Input: A graph $G(V, E)$
Output: Yes, if G is hamiltonian, and No, otherwise.

The supporting evidence for an affirmative answer is just a hamiltonian cycle from G. The spanning character of the cycle can be verified in linear time by testing each of its $|V(G)|$ successive vertices for adjacency.

Hamiltonicity exhibits a common asymmetry: While an affirmative answer to the problem is polynomially verifiable, a negative answer does not seem to be. Thus, while hamiltonicity is in NP, it does not seem to be in co-NP. Despite this, the roles played by yes and no in the decision problem formulation are only nominal. For example, we can always define a *complementary problem* by interchanging the roles of Yes and No in the output to the problem. The decision problem complementary to hamiltonicity is

Problem: Nonhamiltonicity
Input: A graph $G(V, E)$
Output: Yes, if G is not hamiltonian, and No, otherwise.

Nonhamiltonicity is in co-NP since a negative answer is polynomially verifiable: we just provide a hamiltonian cycle from G. Equivalently, nonhamiltonicity is in co-NP, because its complementary problem Hamiltonicity is in NP.

The NP or co-NP status of a problem is not always easy to decide. For example, Primality is in co-NP, since a negative answer can be easily verified by providing a factor. But, Primality is also in NP, though the supporting evidence is based on the following esoteric number-theoretic characterization of primality.

Theorem (Characterization of Primality). A positive integer p is prime if and only if there exists a positive integer n satisfying

(1) $n^{p-1} = 1 \bmod p$.
(2) For every prime divisor x of $p - 1$: $n^{(p-1)/x} <> 1 \bmod p$.

Refer to any number theory text for a proof. This theorem does not lead to a polynomial algorithm for primality. However, it does provide a way to establish polynomial verification of primality using n and the prime factors of $p - 1$. These numbers have total length polynomial in the length of p, and the conditions (1) and (2) take polynomial time to test. For example, there are at most $\lg p$ prime factors of $p - 1$, since $p = 2^{\lg p}$. Therefore, at most $\lg p$ tests of type (2) must be done. Each test requires time polynomial in $\lg p$, if done carefully. For example, each successive power of n required in (2) can be computed in $O(\log^2 p)$ time using repeated squaring. It follows that primality is in NP.

The problem classes NP and co-NP are important because many combinatorial problems are in these classes, and because many classical difficult combinatorial problems satisfy a completeness property (NP-Completeness) with respect to these classes. An NP problem X is said to be NP-Complete if every NP problem can be solved by a polynomial number of calls to a procedure that solves X. The theory of NP and NP-Complete problems occupies a central place in algorithmic graph theory, and is considered in the next section.

Nondeterministic Turing machines. We shall now give a formal definition of the class of problems NP in terms of Nondeterministic Turing Machines of polynomial time performance. Recall that in the case of a DTM a Turing Table defines a single valued mapping or function. If the Turing Table mappings are multivalued, that is, if the table maps pairs in $S \times A$ to sets in $S \times A \times \{0, +1, -1\}$, rather than into unique triples, then the TM is called a *Nondeterministic Turing Machine* (NDTM).

A Nondeterministic Turing Machine can be thought of as a generalization of a sequential Deterministic Turing Machine, obtained by allowing parallel processing of alternatives. Thus, when a Nondeterministic Turing Machine arrives at a point in its execution where multiple alternatives are possible, and which in a DTM would be examined sequentially, the NDTM examines the alternatives in parallel. We can visualize the NDTM as spawning multiple copies of itself, including its current state information and tape data. When the machine is faced with k alternatives, it spawns k copies. Each copy then continues to execute independently on its assigned alternative. If any copy finds a solution to the problem, it reports success and signals all the other copies to halt. The copies neither share data nor communicate otherwise. Some copies may halt without success, and some may not halt at all. A NDTM is said to solve a decision problem Q if some copy halts in the yes or accepting state whenever the input instance is true, but no copy halts in the accepting state when the input instance is false.

Backtracking algorithms, like the hamiltonian path algorithm in Chapter 1, can be interpreted as simulations of NDTMs. For example, at each point where the hamiltonian backtrack algorithm considered k ways to extend a path by appending one of k alternative vertices, which it then examined sequentially, a NDTM algorithm would spawn k copies of the algorithm, one to explore each alternative, the process continuing until a hamiltonian path was found, or shown not to exist.

The *space complexity of a NDTM* is defined just as for a DTM. The *time complexity $T(n)$ of a Nondeterministic Turing Machine* TM, is defined as a function of the length n of the inputs to the TM, and equals the maximum number of steps the NDTM takes to recognize any accepted input of length n, where the time it takes to recognize an accepted input equals the length of the shortest accepting computation on the NDTM for that input. For example, if for some input of length n which is not accepted the TM never halts, but the length of the shortest accepting computation is bounded by $p(n)$ for every accepted input of length n, then $T(n)$ is bounded by $p(n)$. If p is polynomial, the TM is called a *Polynomially (Time) Bounded Nondeterministic Turing Machine* (PNDTM).

The class of problems NP is defined as the collection of all languages L on a finite alphabet A for which there is a polynomially bounded nondeterministic Turing machine. Alternatively, a decision problem Q is in NP if the set of instances of Q with yes answers is precisely the language accepted by some NDTM that has polynomial time complexity. This definition is equivalent to our initial definition of NP as meaning polynomial verifiability for yes answers supported by evidence. For, the verification procedure can be used to trivially define a PNDTM which nondeterministically generates every possible solution to a problem instance, testing each concurrently in nondeterministic polynomial time, and thus accepting in polynomial time.

The following theorem gives an upper bound on the time complexity of a problem in NP.

Theorem (Exponential Time Bound on NP). Let Q be a problem in NP, and let X of length $|X|$ be an instance of Q. Then, X can be solved by a deterministic algorithm in time $O(2^{q(|X|)})$, for some polynomial q.

The proof is as follows. Let $p()$ be the time complexity of the PNDTM for Q that exists by supposition. The number of copies spawned by a given copy of this PNDTM at any point is uniformly bounded, since the numbers of control unit states, tape symbols, and movements are bounded. In particular, if the degree of multivaluedness is at most c, then after n steps the NDTM can contain at most $O(c^n)$ copies executing in parallel. If we take n equal to $p(|X|)$, which is an upper bound on the time to determine that X is accepted, a sequential implementation of this leading part of the nondeterministic algorithm, takes time $O(2^{q(|X|)})$, for a suitable polynomial q, as was to be shown.

The proof of the theorem underscores the computing power of a NDTM. The number of operations implemented can increase exponentially with the run-time of the machine. Because of this phenomenon, the nondeterministic model of computation seems to be more computationally powerful or efficient than the deterministic model. Thus, it is an outstanding conjecture of complexity theory that NP is not equal to P.

The NP-completeness results in the following section represent considerable circumstantial evidence in this direction, but the question is still open.

We can specify a nondeterministic algorithm by defining its Turing Table. We can also define it in terms of the three primitives, Choice, Fail, and Succeed. If S is a set, Choice(S) is a multivalued function of S, which returns the elements of S, and also spawns or creates $|S|$ additional copies of the algorithm, one for each element of S. Each copy then continues to execute the algorithm independently using its particular value from S. The primitive Fail makes the copy executing it halt; while, the primitive Succeed forces all copies to halt. For example, we can define a nondeterministic algorithm for the traveling salesman problem as follows. Let the graph $G(V, E)$ be represented by, say, an adjacency matrix, whose entries contain the weight $w(i, j)$ of each edge (i, j).

```
ND_Procedure ND_TSP (G, Budget, C)

(* Returns spanning cycle C of G of cost ≤ Budget, or fails *)

var  G: Graph
     S: Set of 1..|V(G)|
     k: 0..|V(G)|
     P(|V(G)|), C(|V(G)|): 1..|V(G)|
     Cost, Budget: Real
     ND_TSP, Empty, Succeed, Fail: Boolean function

Set S to {1..|V(G)|}
Set k to 0
Set Cost to 0
```

```
while not Empty(S) do Set k to k + 1
                     Set P(k) to Choice(S)
                     Set S to S − {P(k)}
                     if k > 1
                     then Set Cost to Cost + w(P(k − 1), P(k))

if     Cost + w(P(|V(G)|), P(1)) ≤ Budget
then   Set C to (P(1),...,P(|V|)); Succeed
else   Fail


End_ND_Procedure_ND_TSP
```

Optimization and decision problems. Decision problems are the customary reference point for discussing the complexity of problems, but optimization problems are more common in practice. However, the restriction to decision problems is more apparent than real, for, under very general conditions, the decision and optimization formulations of most problems are polynomially equivalent, as we shall now show.

We will define four types of problems, the first of which is a decision problem called the recognition problem, while the rest are optimization problems, which we will show are polynomially equivalent to the recognition problem. For convenience, we assume the underlying problem is to optimize (or minimize) the value of a polynomial time computable objective function $cost(x, y)$, of a problem instance x and a feasible solution y to the problem instance. The problem types are

Type: Existence of a Parametrically Constrained Solution
Input: Problem instance x, cost function $c(x, y)$, and parameter c
Output: Yes, if there exists solution y satisfying $cost(x, y) \leq c$, and No otherwise.

Type: Value of an Optimal Solution
Input: Problem instance x and cost function $c(x, y)$
Output: Value of an optimal solution

Type: Construction of a Parametrically Constrained Solution
Input: Problem instance x, cost function $c(x, y)$, and parameter c
Output: A solution y satisfying $cost(x, y) \leq c$

Type: Construction of an Optimal Solution
Input: Problem instance x and cost function $c(x, y)$
Output: A minimum cost solution y

For example, we can formulate hamiltonicity as a recognition problem.

Problem: Hamiltonicity Recognition
Input: Graph x, function $cost(x, y) = |V(x)| − |V(y)|$, and parameter c.
Output: Yes, if some cycle y in x satisfies $cost(x, y) \leq c$, and No otherwise.

The problem is tested with the parameter c equal to 0.

The basic results for constructing a solution to the optimization problems using a polynomial number of calls to a solution to the recognition problem are as follows.

(1) The value of an optimal solution can be found by using binary search on a recognition problem parameter.

(2) A parametrically constrained solution can be constructed by solving the recognition problem for a reduced version of the given problem, a technique called *self-reduction*.

(3) An optimal solution can be constructed by applying the procedure of (2) using the optimal value found via (1).

We now illustrate each of these transformations in detail.

The optimal value problem can be solved using the recognition problem provided the optimal value is bounded by $2^{q(|X|)}$, where q is a polynomial and $|X|$ is the length of the problem instance X. Assume, for simplicity, that the value is nonnegative. We start by solving the recognition problem with $c = 2^{q(|X|)} / 2$. Binary search determines the subsequent choices for c. In general, when the optimal value has been restricted to an interval $[a, b]$, we solve the recognition problem again, but for $c = (a + b) / 2$. If the output is yes, the optimal value lies on $[a, (a+b)/2]$; else it lies on $[(a+b)/2, b]$. The procedure takes at most $\log(2^{q(|X|)})$ or $q(|X|)$ steps, which by supposition is a polynomial in $|X|$.

For example, consider the problem of finding the least weight spanning cycle in a weighted graph $G(V, E)$. The value of an optimal solution is bounded above by the sum of the $|V|$ largest edge weights in G, and below by 0, assuming positive weights. Thus, the size of the representation of the upper bound is linear in the size of the input, so that the value of the bound is bounded by $2^{c|X|}$. Therefore, the binary search method can be efficiently applied.

The self-reduction technique for obtaining a parametrically constrained solution using the recognition problem is the same as the approach used in Chapter 1 to define a constructive perfect matching algorithm from an existential perfect matching algorithm. We again illustrate the technique for the shortest spanning cycle problem: construct a shortest spanning cycle on a graph $G(V, E)$ that has cost at most c. Let Recognition (G, c) denote the Boolean result of the corresponding recognition problem, and assume Recognition(G, c) is true. The self-reduction method merely examines the edges of G successively, and eliminates any edges that are not needed to maintain the solvability of the problem at each point. The method requires at most $O(|V|^2)$ calls to Recognition.

repeat

       Select the next edge x in G
       if Recognition (G − x, c) **then Set** G to G − x

until |E(G)| = N

The construction of an optimal solution is now trivial. We merely use method (1) to identify the value of an optimal solution. Then, we apply method (2) to construct a solution by taking the optimal value as the parameter.

10-3 NP-COMPLETE PROBLEMS

We usually think of a problem A as able to be transformed or reduced to a problem B if any algorithm for B can be used to solve A. If the transformation takes polynomial time, then we call it a polynomial transformation or reduction. More formally, a decision problem D_1 is said to be *polynomially reducible* to a decision problem D_2 if there is a polynomial time computable function f mapping the inputs of D_1 to the inputs of D_2 so that an input I_1 for D_1 has answer yes if and only if $f(I_1)$ has answer yes for D_2. Thus, if D_2 can be solved in polynomial time, and D_1 is reducible to D_2, then D_1 can be solved in polynomial time. We solve D_1 by merely converting its inputs to D_2 inputs, and then using the polynomial procedure for D_2. Polynomial reducibility is a transitive relation because polynomials are closed under polynomial composition.

A decision problem D_1 in NP is said to be *NP-Complete* if every problem in NP is polynomially reducible to D_1. If we drop the requirement that D_1 be in NP, then the problem is said to be *NP-Hard*. By transitivity, if D_1 is NP-Complete, and D_1 is polynomially reducible to D_2 in NP, then D_2 is NP-Complete. There are two alternative definitions of NP-Completeness. A decision problem D_1 in NP is *NP-Complete* if every problem D_2 in NP can be solved by an algorithm that makes a polynomial number of calls to a procedure that solves D_1 together with a polynomial number of ordinary computations. The procedure for D_1 is called an *oracle*. Alternatively, a problem D_1 in NP is said to be *NP-Complete* if D_1 is in NP, and the existence of a polynomial algorithm for D_1 implies the existence of a polynomial algorithm for every problem in NP. The three definitions are not known to be strictly equivalent, though the first definition is at least as restrictive as the others. Any differences between the definitions will not affect our current discussions.

The problem of satisfiability from propositional logic was the first problem shown to be NP-Complete (by Cook). Let $X = \{x_1, \ldots, x_n\}$ be a set of Boolean variables. A *literal* is a variable x_i or its logical complement x_i^c. Denote the set of literals by L. A *clause* is a subset of L. Satisfiability is defined as:

Problem: Satisfiability (SAT)

Input: A set of variables X, and a set of clauses based on X.

Output: Yes, if the variables can be assigned truth values (true or false) so that each clause contains at least one true literal, and No otherwise.

We can equivalently define SAT as the decision problem for realizability of expressions of the form

$$(x_1 \lor x_2 \lor \ldots) \quad \textbf{and} \quad (y_1 \lor y_2 \lor \ldots) \quad \textbf{and} \quad \cdots \quad (z_1 \lor z_2 \lor \ldots),$$

where the individual terms $x_i, y_i, \ldots, z_i$ are literals chosen from L. Such expressions are said to be in normal form. For example

$$(x_1 \lor x_2) \quad \textbf{and} \quad (x_1^c \lor x_2^c) \quad \textbf{and} \quad (x_1^c \lor x_2).$$

This expression is true if x_1 is set to false and x_2 is set to true. The expression

$$(x_1 \lor x_2) \quad \textbf{and} \quad (x_1^c \lor x_2^c) \quad \textbf{and} \quad (x_1^c \lor x_2) \quad \textbf{and} \quad (x_1 \lor x_2^c),$$

on the other hand, is not satisfiable. Satisfiablity is trivially in NP, because an assignment of truth values to the variables can be evaluated in linear time. But, there is no known polynomial time algorithm for deciding satisfiability. If the number of variables is N, it seems one must essentially test all 2^N possible settings for the variables to decide realizability. We refer the reader to the references for a proof of the following theorem.

Theorem (Cook's Theorem).   Satisfiability is NP-Complete.

Many graph theoretic problems have been shown to be polynomial reducible from SAT, and hence NP-Complete. Since many of these problems have long been believed to be nonpolynomial, and since the existence of a polynomial solution to any of them implies the existence of a polynomial solution to all of NP, it is generally believed that NP-Complete problems are not polynomially solvable. Thus, practically speaking, once a problem is known to be NP-Complete, we can turn our attention away from a vain search for an efficient, exact solution to the problem, to a more hopeful quest for an approximate or random solution to the problem, or for an exact solution to a special case of the problem.

Before proceeding, it will be useful to observe that the following restricted form of SAT is also NP-Complete.

Problem: k-Satisfiability (k-SAT)

Input: A set of variables X, a fixed positive integer k, and a set of clauses based on X where each clause contains at most k literals.

Output: Yes, if the variables can be assigned truth values so each clause contains at least one true literal, and No otherwise.

This problem is polynomial when k equals 1 or 2, but NP-Complete starting with $k = 3$. Later on, we shall show 3-Satisfiability is polynomially reducible to the minimum coloring problem, and hence the latter is also NP-Complete.

In addition to being a helpful reference point for problem reduction, parametrized NP-Complete problems like k-SAT enable a more finely tuned view of the complexity of a problem. Thus, let m denote the number of clauses, n the number of literals, and k the maximum number of literals per clause. It is a priori possible that SAT, though NP-Complete, might nonetheless be solvable in time, say, $(mn)^k$, since such a formula while polynomial in m and n is still exponential in k. However, the NP-Completeness of k-SAT prohibits this, since it would imply for $k = 3$, the existence of an $(mn)^3$ algorithm for 3-SAT, contrary to the NP-Completeness of 3-SAT. Thus, we can conclude that SAT is not only NP-Complete, but cannot be solved even by algorithms with performance like $(mn)^k$, which are polynomial in the problem parameters other than k (unless P = NP).

Demonstrations of NP-completeness.   We typically show a problem X is NP-Complete by proving a known NP-Complete problem is polynomially reducible to X.

Problem: Complete Subgraph of Size k

Input: A graph G and a positive integer $k < |V(G)|$.

Output: Yes, if G contains a complete subgraph of order k, and No, otherwise.

The direct exhaustive search algorithm for this problem requires examining all subgraphs of order k. Their number is polynomial in $|V(G)|$ but can be exponential in k. Backtracking improves on the basic search algorithm, but exhibits similar exponential behavior. The following theorem proves this problem is NP-Complete, and so probably inherently exponential.

Theorem (NP-completeness of Complete Subgraph).   Satisfiability is polynomially reducible to the complete subgraph problem, and hence the latter is NP-Complete.

The proof is based on modelling an instance of the satisfiability problem by an instance of the complete subgraph problem. Thus, let EXP be a normal form expression with k clauses. We model EXP by a graph G whose order is polynomial in the size of EXP, and then prove EXP is satisfiable if and only if G has a complete subgraph of order k. G is defined as follows. There is a labelled vertex (x, i) for each literal x in the ith clause of EXP, and a pair of vertices (x, i) and (y, j) are connected by an edge if and only if x is not the logical complement of y and i is not equal to j. Observe that if exp is satisfiable, there are literals from each of its k clauses which can be made true simultaneously. Since each such literal is from a distinct clause, the corresponding vertices are adjacent in G, and so form a complete subgraph of order k. Conversely, if G contains a complete subgraph H of order k, then EXP is satisfiable, since each vertex in H arises from a literal from a different clause, and none of the literals are logical complements. Thus, if we set the variables so that every literal is true, then we force EXP to be true. This completes the proof of the theorem.

For example, the model graph for the expression $(x \textbf{ and } x^c)$ consists of the pair of isolated vertices $(x, 1)$ and $(x^c, 2)$. The expression is trivially unsatisfiable and the model graph accordingly contains no complete subgraph of order 2. The expression $(x_1 \bigvee x_2)$ **and** $(x_1^c \bigvee x_2^c)$ **and** $(x_1^c \bigvee x_2)$ **and** $(x_1 \bigvee x_2^c)$ is also unsatisfiable, and so its model graph contains no complete subgraph of order 4. The model graph for the expression $(x_1 \bigvee x_2 \bigvee x_3)$ **and** $(x_1 \bigvee x_2^c \bigvee x_4^c)$ **and** $(x_3^c \bigvee x_4)$ **and** (x_1^c) contains the complete induced subgraph on $(x_2, 1)$, $(x_4^c, 2)$, $(x_3^c, 3)$, $(x_1^c, 4)$. Setting x_2 to True, and x_1, x_3, and x_4 to False, makes the expression true.

The following two problems are closely related to the complete subgraph problem. Recall from Chapter 8 that a *vertex cover* of a graph $G(V, E)$ is a subset of V such that every edge of G is incident with at least one vertex of V, while a *vertex independent set* is a set of mutually nonadjacent vertices. The *size* of each of these sets is its cardinality. These are the decision problems.

Problem: Vertex Cover (VC)

Input: A graph G and a positive integer $k < |V(G)|$.

Output: Yes, if there is a vertex cover of size $\leq k$, and No otherwise.

Problem: Independent Set (IS)

Input: A graph G and a positive integer $k < |V(G)|$.

Output: Yes, if there is an independent set of size $\geq k$, and No otherwise.

Theorem (NP-completeness of Vertex Cover).   Let $G(V, E)$ be a graph. Then, the problem of determining whether G contains a vertex cover of size at most k is NP-Complete.

VC is clearly in NP. We show the complete subgraph problem is polynomially reducible to VC. Thus, let G^c be the complement of G. G^c has a vertex cover of size k if and only if G has a complete subgraph of size $|V(G)| - k$. Since G^c is polynomial time computable from G, it follows that VC is NP-Complete.

Theorem (NP-completeness of Independent Set).   Let $G(V, E)$ be a graph. Then, the problem of determining whether G contains an independent set of size at least k is NP-Complete.

IS is clearly in NP. Observe that X is an independent set if and only if $V - X$ is a vertex cover. It follows immediately that IS is NP-Complete.

The NP-Completeness of the following problem is more difficult to establish.

Problem: k-Colorability

Input: A graph G and a positive integer k (> 2).

Output: Yes, if the Chromatic Number of G equals k; otherwise, No.

We will show the NP-Completeness of a special case.

Theorem (NP-completeness of Three-colorability).   Satisfiability is polynomially reducible to three-colorability, so the latter is NP-Complete.

We shall show how to represent an arbitrary boolean expression X, which is in normal form and has at most three literals per clause, by a graph which is three-colorable if and only if X is satisfiable. Let k equal the number of clauses and let $x_1, \ldots, x_n$ be the variables. We will use the special (sub)graph shown in Figure 10-1. It has the following important property; if it is validly three-colored (with colors 0, 1, and 2), the apex vertex v_4 must be colored with 0 if and only if the input vertices v_1, v_2, and v_3 are colored with 0.

The model graph $G(V, E)$ for the expression X is as follows. For each variable x_i in X, we include a pair of vertices x_i and x_i^c in G, and the edge (x_i, x_i^c). For each clause c in X, we include a copy $S(c)$ of the graph shown in Fig. 10-1, which is hooked up to the variable vertices x_i and x_i^c as follows. For example, given the clause $c = (x_{10}, x_{20}^c, x_2)$, we connect the input lines for $S(c)$ to the vertices for x_{10}, x_{20}^c, and x_2. That is, v_1, v_2, and v_3 in $S(c)$ would just be the vertices in G for the variables x_{10}, x_{20}^c, and x_2. We also include two additional vertices A and B. There is an edge from A to the apex vertex of each clause subgraph. There is also an edge from B to all the variable vertices x_i and x_i^c, and to A.

Input vertices v_1 v_2 v_3

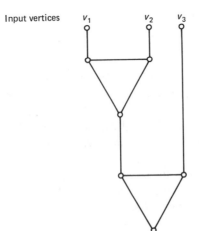

v_4 Apex vertex

Figure 10-1.   Clause subgraph for three-coloring model.

If X is satisfiable, then we can three-color the model graph as follows. Color the vertex for x_i as 1 if x_i is true in the given satisfying assignment for X, and x_i^c as 0. Otherwise, color x_i as 0, and x_i^c as 1. Thus, for every clause, not every literal will be colored 0. Therefore, the apex vertices for every clause can be colored 1 or 2, and so A can be colored 0. Since B is adjacent only to vertices colored 0 or 1, B can be colored 2. This gives a three-coloring of the model graph.

We show next that if the model graph is three-colored, the expression must be satisfiable. Thus, assume that A is colored 0 and B is colored 2. Then, the variable vertices must all be colored either 0 or 1, and the apex vertices must be colored 1 or 2. Therefore, by our previous observation, at least one of the input vertices for every clause cannot be colored 0. Thus, we can make X true by merely taking a literal as true if it is colored 1. In this way, every clause will have at least one true literal. This completes the proof of the theorem.

The proof of the following theorem is by reduction from Vertex Cover, and is quite complicated. However, the subsequent theorem on undirected Hamiltonicity follows readily from this result.

Theorem (NP-completeness of Directed Hamiltonicity). Let $G(V, E)$ be a digraph. Then, the problem of determining whether G contains a directed hamiltonian cycle is NP-Complete.

Theorem (NP-completeness of Undirected Hamiltonicity). Let $G_1(V, E)$ be a graph. Then, the problem of determining whether G_1 contains a hamiltonian cycle is NP-Complete.

The proof is by reduction from directed hamiltonicity. Given a digraph G, we construct an undirected digraph G' from G as follows. For each vertex v in G, we create three vertices v_{in}, v_{out}, and v_{mid} in G' such that; if (u, v) is an edge in G, then (u_{out}, v_{in}) is an edge of G'; if (v, u) is an edge of G, then (v_{out}, u_{in}) is an edge of G'; and (v_{in}, v_{mid}) and (v_{mid}, v_{out}) are edges of G'. G has a directed hamiltonian cycle if and only if G' has a hamiltonian cycle. Since the construction is polynomial time computable, the theorem follows.

List of NP-complete problems. We will present a short list of NP-Complete problems in order to give an idea of their variety. Hamiltonicity is NP-Complete, as are many similar problems such as: the length of a longest path in a graph or the length of a shortest path between a pair of vertices in a digraph in which negative length cycles are allowed. On the other hand, the traveling salesman problem is in NP-Hard. Hamiltonicity is obviously polynomially reducible to the traveling salesman problem, but the latter problem does not appear to be in NP. Compare the difference between the two problems to finding a needle in a haystack (hamiltonicity), as opposed to finding the longest needle in a haystack (traveling salesman).

Sometimes, closely related problems have different complexity status. For example, while we have observed previously that linear programming is polynomial, integer programming is NP-Complete.

Problem: Integer Programming

Input: An integer matrix $\mathbf{A}$ and an integer vector $\mathbf{b}$.

Output: Yes, if there is a 0-1 vector $\mathbf{x}$ such that $\mathbf{Ax} \leq \mathbf{b}$; otherwise, No.

The following inverse problems have different complexities.

Problem: Minimum Cut
Input: A network N, and an integer c.
Output: Yes, if the minimum cut has value $\leq c$ and, no otherwise.

Problem: Maximum Cut
Input: A network N, and an integer C.
Output: Yes, if the maximum cut has value $\geq C$, and no otherwise.

Minimum cut is polynomial, since the Ford and Fulkerson algorithm identifies a minimum cut as a by-product of the maximum flow algorithm in polynomial time. By a *maximum cut,* we mean a cut between some pair of complementary subsets of $V(N)$ of maximum total edge weight. Maximum cut is NP-Complete, even in the restricted case where all the edges have unit weight. However, if the network is planar, then the problem is in P.

We have already shown the following problem is NP-Complete:

Problem: Chromatic Number
Input: A graph G and an integer $k > 2$.
Output: Yes, if the Chromatic Number of $G = k$; otherwise, No.

This remains NP-complete even if G is restricted to be planar. However, for cubic graphs the problem is in P, since by Brook's Theorem the vertex chromatic number is bounded by the maximum degree. The corresponding edge coloring problem, the edge chromatic number problem, is also NP-Complete. Indeed, it remains so even if the graph is restricted to be cubic, whence the only question is, by Vizing's Inequality, whether the edge chromatic number is 3 or 4.

Interestingly, the complexity of graph isomorphism is open.

Problem: Graph Isomorphism
Input: A pair of graphs $G1$ and $G2$.
Output: Yes, if $G1$ is isomorphic to $G2$, and No, otherwise.

This is clearly in NP, since an isomorphism is easily verified, but it is not known whether isomorphism is polynomial or NP-Complete. This remains true for the special cases where G is restricted to be regular or bipartite. However, if the degrees of the graph are bounded by a constant, then the problem is polynomial.

Some typical NP-Complete subgraph problems include the following.

Problem: Bipartite Subgraph
Input: A graph G and integer k.
Output: Yes, if G has a bipartite subgraph with at least k edges, and No otherwise.

Problem: Cubic Subgraph
Input: A graph G and integer k.
Output: Yes, if G has a cubic subgraph with at least k edges, and No otherwise.

The situation varies for pseudopolynomial problems. The following problem is pseudopolynomial.

Problem: Knapsack

Input: A set $\{a_1, \ldots, a_n, b\}$ of positive integers.

Output: Yes, if there is some subset of the a_i's that add up to b, and No otherwise.

Knapsack has pseudopolynomial complexity $O(nb)$, but is NP-Complete. (On the other hand, for example, composite, the complementary problem to primality, is both NP, co-NP, and pseudopolynomial, but is not known to be NP-Complete.)

If a problem remains NP-Complete even when its inputs are encoded in a unary (as opposed to binary) representation, that is, with n represented by n 1s, then the problem is said to be *strongly NP-Complete*. Neither Knapsack, nor any problem with a pseudopolynomial algorithm, is strongly NP-Complete (unless P = NP), because the ordinary (binary) pseudopolynomial algorithm would be polynomial with respect to the unary representation (where the value of the input equals its length).

10-4 RANDOM AND PARALLEL ALGORITHMS

A random algorithm is an algorithm which incorporates probabilistic steps. For example, the probabilistic matching algorithm of Chapter 9, used a random assignment of weights to edges to isolate a unique minimum weight matching. Certain problems that seem to lack efficient deterministic algorithms, have efficient random algorithms. This section introduces the basic complexity classes for random algorithms. We also introduce terminology for describing the performance of parallel algorithms, and complexity classes for parallel algorithms.

Random algorithms.   A decision problem D is said to have *bounded probability of error* if it can be solved in polynomial time by a probabilistic algorithm whose answers are reliable with probability p greater than $1/2$. The class of such problems is denoted by BPP and the associated algorithms are called polynomially bounded *Monte Carlo algorithms*.

The class of *random polynomial problems,* denoted by RP, consists of decision problems which have polynomial time algorithms where a yes answer is always correct, while the probability that an input that should be answered yes triggers a no is less than p, for fixed $p < 1/2$. Thus, even the negative answers are correct with probability at least $1 - p$, that is, more than half the time. Clearly, RP is contained in BPP. The complementary class co-RP is defined similarly.

For example, the Lovasz-Tutte matching algorithm described in Chapter 1 is Random Polynomial. Even though the problem of matching is polynomial, which is presumably a more restrictive problem class than RP, such a random algorithm may be useful because of its simplicity, or faster expected performance.

Problem: Perfect Matching.

Input: A graph $G(V, E)$.

Output: Yes, if G has a complete matching, and No otherwise.

Recall that the Lovasz-Tutte algorithm used the Tutte characterization of graphs with complete matchings, that is, $G(V, E)$ has a complete matching if and only if the Tutte determinant for G is not identically zero. Since the Tutte determinant could not be evaluated in polynomial time, we used a random method of evaluation. When a random instance of the determinant was nonzero, the determinant was definitely nonzero. In our current terminology, this corresponds to a yes answer being always correct. On the other hand, a no answer has a small chance of error, since it might correspond to the fortuitous selection of a root of a nonidentically zero determinant. Thus, the algorithm was of type RP.

We have already remarked on the complexity of primality. We shall now describe a co-RP algorithm for primality.

Theorem (Primality co-RP). The problem Primality is in co-RP.

The proof is as follows. Let n denote a positive integer, and let i be relatively prime less than n. $L(i, n)$ (the *Legendre symbol*) equals $+1$ if $i = x^2$ mod n for some x, and -1 otherwise. $L(i, n)$ can be computed in polynomial time using an algorithm similar to Euclid's algorithm for greatest common divisors. If n is prime, then $L(i, n) = i^{(n-1)/2}$ mod n for every $i < n$; while if n is composite, then $L(i, n) = i^{(n-1)/2}$ mod n for at most half the positive integers less than n which are relatively prime to n. The following procedure uses this to test for primality. Let $GCD(a, b)$ denote the greatest common divisor of a and b.

```
Function Primality (n)

(* Sets Primality to true if n is prime, else false *)

var   i, n: Integer
      Primality: Boolean function

Set i to a random integer less than n

Set Primality to (GCD(i, n) = 1   and*   L(i,n) = i^(n − 1)/2 mod n)

End_Function_Primality
```

The procedure is always correct for no answers, since if i and n are not relatively prime or $L(i, n)$ fails to equal $i^{(n-1)/2}$ mod n, then n is definitely composite. Instances are always correctly answered when n is prime, and a composite is misclassified at most half the time. This does not quite prove the algorithm is co-RP, since the probability of error for a yes answer is not guaranteed to be less than some fixed p less than $1/2$. However, if we merely repeat the procedure t times, classifying n as prime only if the function never fails, then the chance of error is at most $(1/2)^t$. For example, with t equal to 100, the estimated risk of error is less than $(1/2)^{100}$, which is less than the chance of an error due to hardware failure!

The class of *Las Vegas problems,* denoted by LVP, refers to those problems which can be solved by random algorithms whose *expected* time is polynomially bounded, and which never give incorrect answers. That is, for LVP, the answers are always correct, and though the algorithm *may* take exponentially long to terminate, the expected completion time is polynomially bounded. The following theorem gives the relation between LVP and the other probability classes.

Theorem (LVP Characterization). LVP equals the intersection of RP and co-RP.

The proof follows. We will show first that LVP is in RP. Thus, let x be a problem in LVP, and let A_x be an LVP algorithm for x, whose expected time is $q(|x|)$. We can obtain an RP algorithm A_x' for x by running A_x for (say) time $10q(|x|)$, or until it terminates, whichever comes first, and interpreting the results as follows. If an answer is output at any point, we take that as the output for A_x', otherwise, if A_x does not terminate within the time limit, then A_x' outputs No as the answer. The probability is at least 90% that A_x' terminates, in which case the output is definitely correct. On the other hand, some of the default No answers may be incorrect. Thus, A_x' is an RP algorithm for x, as was to be shown. A similar argument, with Yes replacing No, proves LVP is in co-RP.

It remains to prove that RP $\cap$ co-RP is in LVP. The argument is as follows. Let x be a problem in the intersection. Let A_{RP} and $A_{\mathrm{co\text{-}RP}}$ be the associated algorithms for RP and co-RP. If we run A_{RP} and $A_{\mathrm{co\text{-}RP}}$ on x, each taking polynomial time (say) $q(|x|)$, a Yes output from A_{RP} is definitely correct, and a No output from $A_{\mathrm{co\text{-}RP}}$ is definitely correct, and in either case we stop. Ambiguity only arises when A_{RP} outputs No and $A_{\mathrm{co\text{-}RP}}$ outputs Yes. The probability that this occurs is at most p. Therefore, the probability that this happens in k successive (and independent) trials of A_{RP} and $A_{\mathrm{co\text{-}RP}}$ is therefore, at most p^k. Therefore, the expected number of iterations of the algorithm until a correct answer is obtained is the sum of $(1 - p)ip^{i-1}$, over $i \geq 1$, which is bounded by a constant c, so that the expected completion time of the iterated algorithm is bounded by $cq(|X|)$, as required for an LVP algorithm. This completes the proof of the theorem.

As an example of a problem in LVP, consider the *ringleader selection problem*. A collection of processors $p_1, \ldots, p_n$ is distributed in a ring or cycle, with processor p_i connected to processors p_{i+1} and p_{i-1}, and p_n to p_1. The objective is to identify a single processor that all the processors agree will act as a leader processor. Only the distributed communications ability of the ring are allowed in implementing the election. The processors are synchronized, and each processor knows the number of processors in the ring, as well as its predecessor and successor processors. The election algorithm is as follows. Initially, each processor sets a local counter m to n, and a local status flag to active. In each subsequent phase, each processor randomly, uniformly, and independently selects an integer from $1 . . m$. Processors that randomly select 1, announce that fact to the other processors by sending an appropriate message around the ring. In this way, every processor can determine the number of processors k that selected 1. If $k = 1$, then the leader election protocol stops, and the sole processor that selected 1 becomes the leader. If $k = 0$, then there is no change in the status of the system and the whole step is repeated. If $k > 1$, then any processor that selected a value other than 1, sets its status flag to inactive, and does not participate in further steps, except to pass messages; the remaining k processors, that selected 1, initiate the next step, by setting m to k, and once again select a random number on $1 . . m$. We leave it as an exercise to prove that the expected number of steps before the procedure terminates is bounded by a constant, independent of n. Since each step involves $O(n)$ message passing operations, the overall algorithm has expected completion time linear in n, and so is in LVP.

Parallel complexity. The complexity of a sequential algorithm is usually described in terms of its space and time performance. For parallel algorithms, several other measures of performance are also used. Thus, let X be a problem for which the fastest sequential algorithm A_{seq} takes time T_{seq}, while a parallel algorithm A_p with p processors takes time T_p. The *speedup* of the parallel algorithm is defined as

$$T_{\text{seq}} / T_{\text{p}} .$$

For example, if the parallel algorithm is 10 times faster, then the speedup is 10. The *cost* of the parallel algorithm is defined as

$$p \, T_{\text{p}} ,$$

where p may be 1, in which case the cost is just the time. The *efficiency* or *processor utilization* of the parallel algorithm is defined as

$$\text{cost of } A_{\text{seq}} / \text{ cost of } A_p ,$$

which is equivalent to

$$T_{\text{seq}} / (pT_p) .$$

In other words, the efficiency is the ratio of the actual speedup $T_{\text{seq}} / T_{\text{p}}$ to the maximum speedup p. Of course, since any parallel algorithm on p processors with time T_p, can be converted into a sequential algorithm with time pT_p, the efficiency of a parallel algorithm can never be greater than 1. An important performance measure for VLSI circuits is the area occupied by the circuit. See Ullman (1984) for an extensive discussion of area-time tradeoffs and lower bounds.

A fundamental complexity class for parallel algorithms is the class of problems *NC(k)* which have fast parallel algorithms. Problems in this class of size n are solvable in polylog time, that is, in time $O(\log^k(n))$ by parallel algorithms using a number of processors at most polynomial in n. If k is arbitrary, the class is denoted by *NC*. *RNC(k)* and *RNC* denote the class of problems with probabilistic polylog parallel algorithms.

There are some reservations about the lack of specificity of theoretical models of parallel computation such as the shared memory model (see Cook [1983]). An alternative, graph-theoretic model is the Boolean circuit model. A *Boolean circuit* is defined as a labelled acyclic digraph, whose vertices of indegree 0 are considered as input vertices, while its vertices of outdegree 0 are output vertices. In general, the indegree and outdegree of vertices can be arbitrary. If each noninput vertex represents the logical gates **and, or**, etc, the indegree is at most two. A circuit can be conveniently represented in terms of a topological ordering of its vertices as a sequence of instructions. Graph-theoretic techniques are generally useful in the analysis of such circuits. For example, Hoover, Klawe, and Pippenger (1984) use Huffman tree techniques to reduce the outdegrees of a circuit.

The inputs to a circuit are vectors of 0s and 1s. In analogy to a TM that accepts or fails to accept a language, a circuit can be thought of as defining a mapping from an input bit vector of length n to $\{0, 1\}$ ($\{\text{false}, \text{true}\}$). That is, it accepts or rejects an input string, similar to a TM. If A^n denotes a set of bit vectors of length n, then we say a circuit C *realizes* A^n if and only if C accepts precisely the vectors in A^n. The performance of a circuit is characterized by the number of its noninput vertices, which is called the *size* of the circuit, and is a measure of the amount of hardware used, and the length of its longest path, which is called the depth of the circuit, and corresponds to the parallel computation time. $\text{Size}_A(n)$ and $\text{Depth}_A(n)$ are defined as the minimum size and depth of a circuit that realizes A^n.

REFERENCES AND FURTHER READING

Cook (1983) gives an overview of computational complexity from the viewpoint of a founder of the subject. The ground-breaking work on the theory of NP-Completeness was done by Cook (1971) and Karp (1972). Garey and Johnson (1979) is the classic comprehensive work on NP-Completeness. Aho, Hopcroft, Ullman (1974) is an excellent treatment and covers different complexity classes in detail. Our discussion of the relation between decision and optimization problems, and some of the discussion of random complexity, is based on Melhorn (1984). The nondeterministic knapsack procedure is from Reingold, et al. (1977). Brassard and Bratley (1988) give a very interesting treatment of complexity, and many probabilistic algorithms. See also Papadimitriou and Steiglitz (1982).

The NP status of primality is from Pratt (1975). See Brassard (1988) for applications of primality to cryptology. For the complexity of linear programming, refer to Karmarkar (1984) and Khachian (1980). The proof of the complexity of three-coloring is based on Gibbons (1985). Hadlock (1975) gives a polynomial algorithm for maximum planar cut. Thompson (1979) develops the notion of area-time complexity tradeoffs. These questions are considered in great detail in Ullman (1984). See Lint and Agerwala (1981) for a discussion of communication costs in parallel algorithms. For Boolean circuit models of computation and comparisons between different models of computation, see Borodin (1977), Pippenger and Fischer (1979) and Cook (1983) for further references. See also Hoover, Klawe, and Pippenger (1984). The random distributed ringleader exercise is based on Itai and Rodeh (1981). Vergis, Steiglitz, and Dickinson (1986) discuss NP-Completeness in the context of analog systems.

EXERCISES

1. Give a polynomial algorithm for finding a minimum vertex independent set which is also a vertex cover.

2. Let $\{a_i \mid i = 1, \ldots, n\}$ be a subset of positive integers and let N be an integer. Write an $O(n)$ nondeterministic algorithm to determine whether any subset of the a_i's sums to N.

3. Suppose one algorithm takes 2^n steps for a problem of size n, while another algorithm for the same problem, takes n^k steps. For each k, $k = 1, \ldots, 7$, what is the tradeoff point between the performances of the two algorithms, that is, the point where the speed of one algorithm surpasses the speed of the other, as a function of the problem size?

4. Suppose P(E) is a procedure that can test whether a normal form logical expression E in n variables $x_1, \ldots, x_n$ is satisfiable. Show how to use P to determine an assignment of truth values to the x_i that makes E true in n calls to P (provided E is satisfiable).

5. Prove Knapsack is pseudopolynomial.

6. Prove 2-SAT is polynomial.

7. If an inappropriate traversal technique is used in the Ford and Fulkerson maximum flow algorithm, the number of iterations may be as large as the capacity of a minimum capacity cut. What is the complexity of the algorithm under these circumstances?

8. Design and establish the performance of a polynomial algorithm for computing the integer square root of a positive integer.

9. Consider the following greedy maximum vertex independent set algorithm. For each vertex v, add v to the independent set if it is not adjacent to any vertex already in the independent set. Show that the relative error of this algorithm can be arbitrarily large.

10. Prove the following problem is NP-Complete. Use the NP-Completeness of the complete subgraph problem.

    Problem: Set Packing

    Input: A family F of sets and a positive integer k.

    Output: Yes, if F contains k pairwise disjoint sets; otherwise, No.

11. Prove the following problem is NP-Complete.

    Problem: Longest Path

    Input: A graph $G(V, E)$, a pair of vertices x and y in $V(G)$, and a positive integer k.

    Output: Yes, if G contains a path from x to y of length at least k; otherwise, No.

12. Prove the following restricted version of the hamiltonian path problem is NP-Complete.

    Problem: Bipartite Hamiltonicity

    Input: A bipartite graph $G(V, E)$.

    Output: Yes, if G contains a hamiltonian path; otherwise, No.

13. Prove the following problem is NP-Complete.

    Problem: Degree Constrained Spanning Tree

    Input: A graph $G(V, E)$ and a positive integer k.

    Output: Yes, if G contains a spanning tree of maximum degree k; otherwise, No.

14. Prove the following problem is NP-Complete. (Try reduction from the chromatic number problem.)

    Problem: Clique Cover

    Input: A graph $G(V, E)$, and an integer k.

    Output: Yes, if $V(G)$ is the union of at most k disjoint sets V_i, such that the induced subgraphs on the V_i are complete; otherwise, No.

15. Prove the maximum cut problem is NP-Hard.

16. Prove the following problem is NP-Complete.

    Problem: Steiner Tree

    Input: A graph $G(V, E)$, a real number w, a weight function defined on $E(G)$, and a subset V' of V.

    Output: Yes, if G contains a tree spanning V' and, if necessary, some other vertices of V, of weight at most w; otherwise, No.

17. Prove that the expected number of election phases in the ringleader election problem is bounded by a constant.

Bibliography

AGGARWAL, A., and R. ANDERSON (1987), "A random NC algorithm for depth first search," in *Proc. 19th Ann. ACM Sympos. on Theory of Computing*, ACM, Baltimore, MD, 325-334.

AHO, A. V., M. R. GAREY, and J. D. ULLMAN (1972), "The transitive reduction of a directed graph," *SIAM Jour. Computing*, 1, 131-137.

AHO, A. V., J. E. HOPCROFT, and J. D. ULLMAN (1974), *The Design and Analysis of Computer Algorithms*, Addison-Wesley, Reading, MA.

AKL, S. G. (1985), *Parallel Sorting Algorithms*, Academic Press, Inc, New York.

APPEL, K., and W. HAKEN (1977), "The solution to the four-color problem," *Scientific American*, 27(4), 108-121.

BANACHOWSKI, L. (1980), "A complement to Tarjan's result about the lower bound on the complexity of the set union problem," *Information Processing Letters*, 11, 59-65.

BASSE, S. (1988), *Computer Algorithms: Introduction to Design and Analysis* (2nd ed.), Addison-Wesley, Reading, MA.

BECKER, M., et al. (1982), "A probabilistic algorithm for vertex connectivity of graphs," *Information Processing Letters*, 15(3), 135-136.

BEHZAD, M., G. CHARTRAND, and L. LESNIAK-FOSTER (1979), *Graphs and Digraphs*, Wadsworth, Belmont, CA.

BENTLEY, J. L. (1980), "A parallel algorithm for constructing minimum spanning trees," *Jour. Algor.*, 1(1), 51-59.

BHABANI, P. SINHA, et al. (1986), "A parallel algorithm to compute the shortest paths and diameter of a graph and its VLSI implementation," *IEEE Trans. Comput.*, C-35, 11, 1000-1004.

BOESCH, F., ed. (1976), *Large-Scale Networks: Theory and Design*, IEEE Press, New York.

BOESCH, F. (1982), "Introduction to basic network problems," in *The Mathematics of Networks* (S. Burr, ed.), *Proc. of Symposia in Applied Math.*, Amer. Math. Soc., 26, Providence, 1-29.

BOLLOBAS, B., ed. (1982), *Graph Theory*, North Holland Mathematical Studies, No. 62, New York.

BONDY, J. A., and U. S. R. MURTY (1976), *Graph Theory with Applications*, The Macmillan Press, London.

BORODIN, A. (1977), "On relating time and space to size and depth," *SIAM Jour. Computing*, 6, 733-744.

BRASSARD, G., and P. BRATLEY (1988), *Algorithmics Theory and Practice,* Prentice-Hall, Englewood Cliffs, NJ.

BRASSARD, G. (1988), *Modern Cryptology: A Tutorial,* Lecture Notes in Computer Science, Springer-Verlag, New York.

BUSACKER, R. G., and T. L. SAATY (1965), *Finite Graphs and Networks: An Introduction with Applications,* McGraw-Hill, New York.

CAPOBIANCO, M., and J. MOLLUZZO (1978), *Examples and Counterexamples in Graph Theory,* North-Holland, New York.

CHIBA, N., T. NISHIZEKI, and N. SAITO, (1981), "A linear 5-coloring algorithm of planar graphs," *Jour. Algor.,* 2(4), 317-327.

CHRISTOFIDES, N. (1975), *Graph Theory: An Algorithmic Approach,* Academic Press, New York.

CHRISTOFIDES, N. (1976), "Worst case analysis of a new heuristic for the travelling salesman problem," *Tech. Rep.,* Grad. School of Industrial Administration, Carnegie-Mellon, Pittsburgh, PA.

CLARKSON, K. L. (1987), "Approximation algorithms for shortest path motion planning," in *Proc. 19th Ann. ACM Sympos. on Theory of Computing,* ACM, Baltimore, MD, 56-65.

COFFMAN, E. G., and P. J. DENNING (1973), *Operating Systems Theory.* Prentice-Hall, Englewood Cliffs, NJ.

COOK, S. A. (1971), "The complexity of theorem-proving procedures," *Proc. 3rd Annual Sympos. on Theory of Computing,* 151-158.

COOK, S. A. (1983), "An overview of computational complexity," *Commun. ACM,* 26(6), 401-408.

DEO, N. (1974), *Graph Theory with Applications to Engineering and Computer Science,* Prentice-Hall, Englewood Cliffs, NJ.

DEMOUCRON, G., Y. MALGRANGE, and R. PERTUISET (1964), "Graphes planaires: reconnaissance et construction de representations planaires topologiques," *Rev. Francaise Recherche Operationnelle,* 8, 33-47.

DEVROYE, L. (1986) "A note on the average height of binary search trees," *Jour. ACM,* 33(3), 489-498.

DIJKSTRA, E. (1959), "Two problems in connexion with graphs," *Num. Math.,* 1, 269-271.

DOLEV, D. (1982), "The Byzantine generals strike again," *Jour. Algor.,* 3(1), 14-30.

EDMONDS, J. (1965), "Paths, trees, and flowers," *Canadian Jour. Math.,* 17 (4), 449-467.

EDMONDS, J., and D. R. FULKERSON (1968), "Bottleneck extrema," Memorandum RM-5375-PR, RAND Corp., Santa Monica, CA.

EDMONDS, J., and R. M. KARP (1972), "Theoretical improvements in algorithmic efficiency for network flow problems," *Jour. ACM.,* 19, 248-264.

ERDOS, P., and A, RENYI (1960), "On the evolution of random graphs," *Magyar Tud. Akad. Mat. Kut. Int. Kozl,* 5, 17-61.

EVEN, S. (1979), *Graph Algorithms,* Computer Science Press, Potomac, MD.

FLAJOLET, P., and A. ODLYZKO (1982), "The average height of binary trees and other simple trees," *Jour. of Comp. and System Science,* 25(2), 171-213.

FORD, L. R., and D. R. FULKERSON (1962), *Flows in Networks,* Princeton Univ. Press, Princeton, NJ.

FREDMAN, M. L., and R. E. TARJAN (1987), "Fibonacci heaps and their use in improved network optimization algorithms," *Jour. ACM,* 34(3), 596-615.

GABOW, H. N.(1976), "An efficient implementation of Edmonds' algorithm for maximum matching on graphs," *Jour. ACM,* 23, 221-234.

GABOW, H. N., and R. E. TARJAN (1983), "A linear-time algorithm for a special case of disjoint set union," *Proc. 15th Ann. ACM Sympos. on Theory of Computing,* 246-251.

GAREY, M., and D. JOHNSON (1979), *Computers and Intractability: A Guide to the Theory of NP-Completeness.* Freeman and Co., San Francisco.

GIBBONS, A. (1985), *Algorithmic Graph Theory,* Cambridge Univ. Press, London.

GONDRAN, M., and M. MINOUX (1984), *Graphs and Algorithms,* John Wiley and Sons, New York.

GOTLIEB, C. C., and L. R. GOTLIEB (1978), *Data Types and Structures,* Prentice-Hall, Englewood Cliffs, NJ.

GRAHAM, R. L., and P. HELL (1985), "On the history of the minimum spanning tree problem," *Annals Hist. Computing,* 7(1), 43-57.

HADLOCK, F. O. (1975), "Finding a maximum cut of a planar graph in polynomial time," *SIAM Jour. Computing,* 4, 221-225.

HARARY, F. (1971), *Graph Theory,* Addison-Wesley, Reading, MA.

HOPCROFT, J. E., and R. E. TARJAN (1974), "Efficient planarity testing," *Jour. ACM,* 21, 549-568.

HOOVER, H. J., M. M. KLAWE, and N. J. PIPPENGER (1984), "Bounding fanout in logical networks," *Jour. ACM.,* 31(1), 13-18.

HOROWITZ, E., and S. SAHNI (1978), *Fundamentals of Computer Algorithms,* Computer Science Press, Potomac, MD.

HU, T. C. (1970), *Integer Programming and Network Flows,* Addison-Wesley, Reading, MA.

HU, T. C. (1982), *Combinatorial Algorithms,* Addison-Wesley, Reading, MA.

HWANG, K., and F. A. BRIGGS (1984), *Computer Architecture and Parallel Processing,* McGraw-Hill, New York.

ITAI, A., and M. RODEH (1981), "Symmetry breaking in distributive networks," in *Proc. 22nd Ann. IEEE Sympos. on Foundations of Computer Science,* 150-158.

ITALIANO, G. F., (1986), "Transitive closure for dynamic digraphs," Univ. di Roma, *La Sapienza,* Dipart. Informat. e Sistemist., Roma, Italy.

JOHNSON, D. S., C. ARAGON, L. MCGEOCH, and C. SCHEVON (1988), "Optimization by simulated annealing (part I)," to be published.

KAMEDA, T. (1980), "Testing deadlock freedom of computer systems," *Jour. ACM,* 27(2), 270-280.

KARMARKAR, N. (1984), "A new polynomial-time algorithm for linear programming," *Combinatorica,* 4(4), 373-395.

KARP, R. (1972), "Reducibility among combinatorial problems," in *Complexity of Computer Computations,* R. E. Miller and J. W. Thatcher, eds., Plenum Press, New York, NY, 85-104.

KARP, R. M., and TARJAN, R. E. (1980), "Linear expected-time algorithms for connectivity problems," *Jour. Algor.,* 1 (4), 374-393.

KARP, R. M., and A. WIGDERSON (1984), "A fast parallel algorithm for the maximal independent set problem," in *Proc. of 16th ACM Sympos. on Theory of Computing,* 266-272.

KERNIGHAN, B. W. (1971), "Optimal sequential partitions of graphs," *Jour. ACM,* 18 (1), 34-40.

KHACHIAN, L. G. (1980), "Polynomial algorithms in linear programming," *Zhurnal Vychislitelnoi Mathematiki i Matematicheskoi Fiziki,* 20, 53-72.

KNUTH, D. E. (1968), *The Art of Computer Programming Vol. 1: Fundamental Algorithms,* Addison-Wesley, Reading, MA; 2nd ed., 1973.

KNUTH, D. E. (1969), *The Art of Computer Programming Vol. 2: Seminumerical Algorithms,* Addison-Wesley, Reading, MA; 2nd ed., 1981.

KNUTH, D. E. (1971), "Optimum binary search trees," *Acta Informatica*, 1(1), 14-25.

KNUTH, D. E. (1973), *The Art of Computer Programming Vol. 3: Sorting and Searching*, Addison-Wesley, Reading, MA.

KNUTH, D. E., and F. R. STEVENSON (1973), "Optimal measurement points for program frequency counts," *BIT*, 13(3), 313-322.

KRUSKAL, J. B. (1956), "On the shortest spanning subtree of a graph and the traveling salesman problem," *Proc. Amer. Math. Soc.*, 7, 48-50.

KUNG, H. T. (1980), "The structure of parallel algorithms," in *Advances in Computers* Vol. 19 (M. Yovits, ed.), Academic Press, New York, 65-111.

KUNG, H. T. (1982), "Why systolic architectures?" *Computer*, 15(1), 37-46.

LAKSHMIVARAHAN, S., et al. (1984), "Parallel sorting algorithms," in *Advances in Computers*, Vol. 23 (M. Yovits, ed.), Academic Press, New York, 295-354.

LANG, H., M. SCHIMMLER, H. SCHMECK, and H. SCHRODER (1985) "Systolic sorting on a mesh connected network," *IEEE Trans. Computing*, C-34, 652-658.

LAWLER, E. L. (1976), "Graphical algorithms and their complexity," in *Foundations of Computer Science II, Part 1*, Mathematisch Centrum, Amsterdam, 3-32.

LAWLER, E. L. (1976), *Combinatorial Optimization: Networks and Matroids*, Holt, Rinehart and Winston, New York.

LAWLER, E. L., J. K. LENSTRA, A. H. G. RINOOY KAN, and D. B. SHMOYS, eds. (1984), *The Traveling Salesman Problem*, John Wiley, New York.

LIN, S. (1965), "Computer solution of the traveling salesman problem," *Bell Syst. Tech. Jour.*, 44, 2245-2269.

LINT, B., and T. AGERWALA (1981), "Communication issues in the design and analysis of parallel algorithms," *IEEE Trans. Software Engrg.*, SE-7, 174-188.

LIPTON, R., and TARJAN, R. E. (1977) "Applications of a planar separator theorem," in *18th IEEE Sympos. on Foundations of Computer Science*, 162-170.

LOVASZ, L. (1979), "On determinants, matchings, and random algorithms," in *Fundamentals of Computing Theory* (L. Budach, ed.), Akademia-Verlag, Berlin.

LOVASZ, L., and M. PLUMMER (1987), *Matching Theory*, Academic Press, Budapest, Hungary.

LUBY, M. (1986), "A simple parallel algorithm for the maximal independent set problem," *SIAM Jour. Computing*, 15(4), 1036-1053.

MALHOTRA, V. M., M. PRAMODH KUMAR, and S. N. MAHESHWARI (1978), "An $O(|V|^3)$ algorithm for finding maximum flows in networks," Computer Science Program, Indian Institute of Technology, Kanpur, India.

McHugh, J. A. (1984), "Hu's precedence tree scheduling algorithm: a simple proof," *Naval Res. Logist. Quart.*, 31, 409-411.

McHugh, J. A. (1986), "Data structures," in *Handbook of Modern Electronics and Electrical Engineering*, John-Wiley, New York, 2062-2079.

MEAD, C., and L. CONWAY (1980), *Introduction to VLSI Systems*, Addison-Wesley, Reading, MA.

MEHLHORN, K. (1975), "Nearly optimal binary search trees," *Acta Informatica*, 5(4), 287-295.

MEHLHORN, K. (1984), *Data Structures and Algorithms 2: Graph Algorithms and NP-Completeness*, Springer-Verlag.

MOLODOVAN, D. I. (1983), "On the design of algorithms for VLSI systolic arrays," *Proc. IEEE*, 71(1), 113-120.

MULMULEY, K., U. V. VAZIRANI, and V. V. VAZIRANI (1987), "Matching is as easy as matrix inversion," in *Proc. 19th Ann. ACM Sympos. on Theory of Computing*, ACM, Baltimore, MD, 345-354.

NORMAN, R. Z., and M. O. RABIN (1959), "An algorithm for a minimum cover of a graph," *Proc. Amer. Soc.*, 10, 315-319.

PAIGE, R. C., and C. P. KRUSKAL (1985), "Parallel algorithms for shortest paths problems," *Proc. of 1985 Internat. Conf. on Parallel Processing*, Pennsylvania State Univ., University Park, PA, 14-20.

PAN, V. (1985), "Fast and efficient algorithms for the exact inversion of integer matrices," *Fifth Ann. Foundations of Software Technology and Theoretical Computer Science Conference.*

PAPADIMITRIOU, C. H., and K. STEIGLITZ (1982), *Combinatorial Optimization: Algorithms and Complexity*, Prentice-Hall, Englewoods Cliffs, NJ.

PETERSON, J. L., and A. SILBERSCHATZ (1985), *Operating System Concepts* (2nd ed.), Addison-Wesley, Reading, MA.

PIPPENGER, N. J., and FISCHER, M. J. (1979), "Relations among complexity measures," *Jour. ACM.*, 26(2), 361-381.

POLYA, G. (1957), *How to Solve It*, Doubleday Anchor, Garden City, NY.

POLYA, G. (1968), *Mathematical Discovery: On Understanding, Learning, and Teaching Problem Solving*, Wiley, NY.

PRATT, V. R. (1975), "Every prime has a succinct certificate," *SIAM Jour. Computing*, 4(3), 214-220.

PRIM, R. C. (1957), "Shortest connection networks and some generalizations," *Bell Syst. Tech. Jour.*, 36, 1389-1401.

PURDOM, P. W. and C. A. BROWN (1985), *The Analysis of Algorithms*, Holt, Rinehart, and Winston, New York.

QUINN, M. J., and N. DEO (1984), "Parallel graph algorithms," *ACM Comput. Surveys*, 16(3), 319-348.

REINGOLD, E. M., J. NIEVERGELT, and N. DEO (1977), *Combinatorial Algorithms: Theory and Practice*, Prentice-Hall, Englewood Cliffs, NJ.

ROBERTS, F. S. (1984), *Applied Combinatorics*, Prentice-Hall, Englewood Cliffs, NJ.

SEDGEWICK, R. (1983), *Algorithms*, Addison-Wesley, Reading, MA.

SEDGEWICK, R., and J. S. VITTER (1986), "Shortest paths in Euclidean graphs," *Algorithmica*, 1(1), 31-48.

SETHI, R. (1975), "Complete register allocation problems," *SIAM Jour. Computing*, 4(3), 226-248.

SETHI, R., and J. D. ULLMAN (1970), "The generation of optimal code for arithmetic expressions," *Jour. ACM*, 17(4), 715-728.

SHAMOS, P. M. (1975), "Geometric complexity," in *Seventh Symposium on Theory of Computing*, ACM, New York.

SHAMOS, M. I., and D. HOEY (1975), "Closest-point problems," in *Sixteenth Ann. Sympos. Foundations Computer Science*, IEEE Comp. Soc., Piscataway, NJ, 151-162.

SHILOACH, Y. and U. VISHKIN (1982), "An O(log n) parallel connectivity algorithm," *Jour. of Algor.*, 3, 57-67.

SKISCIM, C. C., and B. L. GOLDEN (1987), "Computing k-shortest path lengths in euclidean networks," *Networks*, 17, 341-352.

SLEATOR, D. D., and R. E. TARJAN (1981), "A data structure for dynamic trees," *Proc. 13th Ann. ACM Sympos. on Theory of Computing,* ACM, New York, 114-122.

STANDISH, T. A. (1980), *Data Structure Techniques,* Addison-Wesley, Reading, MA.

STONE, H. S. (1971), "Parallel processing with the perfect shuffle," *IEEE Trans. Computing,* C-20, 153-161.

STONE, H. S. (1977), "Multiprocessor scheduling with the aid of network flow algorithms," *IEEE Trans. Software Engrg.,* 3(1), 237-245.

SWAMY, M. N. S., and K. THULASIRAMAN (1981), *Graphs, Networks, and Algorithms.* John Wiley & Sons, New York.

TARJAN, R. E. (1972), "Depth first search and linear graph algorithms," *SIAM Jour. Computing,* 1, 146-160.

TARJAN, R. E. (1974), "Finding dominators in directed graphs," *SIAM. Jour. Computing,* 3, 62-89.

TARJAN, R. E. (1979), "A class of algorithms which require nonlinear time to maintain disjoint sets," *Jour. Comput. Syst. Science,* 18, 110-127.

TARJAN, R. E. (1983), *Data Structures and Network Algorithms,* Soc. Indust. Appl. Math., Philadelphia.

TARJAN, R. E., and J. VAN LEEUWEN (1984) "Worst-case analysis of set union algorithms," *Jour. ACM.,* 31(2), 245-281.

THOMPSON, C. D. (1979), "Area-time complexity for VLSI," *Proc. 11th Ann. ACM Sympos. Theory of Computing,* Atlanta, GA, (April-May), 81-88.

TURAN, P. (1954), "On The Theory of Graphs," *Colloq. Math.,* 3, 19-34.

TUTTE, W. T. (1963), "How to draw a graph," *Proc. London Math. Soc.,* 13 (3), 743-767.

ULLMAN, J. D. (1984), *Computational Aspects of VLSI,* Computer Science Press, Potomac, MD.

VALIANT, L. G. (1975), "Parallelism in comparison problems," *SIAM Jour. Computing,* 4 (3), 348-355.

VALIANT, L. G., and G. J. BREBNER (1981), "Universal schemes for parallel communication," *Proc. 13th Ann. ACM Sympos. on the Theory of Computing,* 263-277, ACM, New York.

VERGIS, A., K. STEIGLITZ, and B. DICKINSON (1986), "The complexity of analog computation," *Math. and Comp. in Simulat.,* 28, 91-113.

VISHKIN, U. (1983), "Implementation of simultaneous memory access in models that forbid it," *Jour. Algor.,* 4, 45-50.

WELSH, D. J. A. (1983), "Randomized algorithms," *Disc. Appl. Math.,* 5(1), 133-145.

WIGDERSON, AVI. (1983), "Improving the performance guarantee for approximate graph coloring," *Jour. ACM.,* 30(4), 729-735.

WILLIAMSON, S. G. (1984), "Depth-first search and Kuratowski subgraphs," *Jour. ACM,* 31 (4), 681-693.

WILSON, R. J. (1985), *Introduction to Graph Theory,* Longman, Harlow, Essex, UK.

WONG, C. K. (1983), *Algorithmic Studies in Mass Storage Systems,* Computer Science Press, Potomac, MD.

YAO, A. (1975), "An $O(|E| \log \log |V|)$ algorithm for finding minimum spanning trees," *Information Processing Letter,* 4(1), 21-23.

Index

O

Odd cycles, 15, 248–49
Odlyzko, 119
Optimal file merging, 123
Optimal program segmentation,
   58–61
Optimal record layout, 70–71
Optimal register allocation, 141–45
Optimal schedules, 45
Optimal search trees, nearly, 71–72
Optimization and decision
   problems, 301
Or*, 48, 188
Order, 1, 8
Order, H(v), 268
Ordered tree, 116
Ore's hamiltonicity condition, 45
Orientable graph, 163
Orientation, 163
Out-degree, 3
Outer vertices, 22, 24

P

Pairwise connectivity, 212
Pairwise disconnecting set, 212
Pan, 274
Parallel algorithms, 256–93
   complexity, 312–13
   connected components, 265–70
   cost, 312
   Dijkstra's shortest path, 262
   efficiency, 312
   Floyd's shortest paths, 264
   heap creation, 278
   heap deletion, 275
   hypercube matrix multiplication,
      283–90
   maximal independent set, 292–93
   maximum, 240–42
   maximum matching, 270–74
   minimum, 263
   minimum spanning tree, on tree
      processor, 292
   processor utilization, 312
   ringleader selection, 311
   speedup, 312
Parent:
   of tree vertex, 115
   in unary tree, 130
Partitioning, 55
   for shortest paths, 55–56
Part (of graph with respect to planar
   representation), 34
Parts explosion, 136
Path, 2, 4
Path compression, 133
Path length, 118
Path optimality, 100–103

Path representation in Floyd's
   algorithm, 98–99
Paths, longest, 140
Paths, shortest: *See* Shortest path
Pebble game, 142–43
Pending edge, 34
Perfect graph, 53
Perfect matching, 14
Permutation, 28, 272,
PERT, 140–41
Petersen, 239
Petersen graph, 19, 36–37, 54, 221
Planar dual, 32
Planar embedding, 31
Planar extendible, 34
Planar extension, 34
Planar graph, 31, 38–39
   regular, 53
Pointer graph, 268
Polynomially reducible, 303
Polynomially time bounded
   nondeterministic Turing
   machine, 299
Polynomial problems, 294–97
   closure, 294
Polynomial time verifiable, 297
Positional pointer, 13–14
Potential, 110–11
Potentially drawable, 35
Power (of graph), 3
Powers of adjacency matrix, 9
(p, q) graph, 2
PRAM (parallel random access
   memory), 262
Precedence digraph, 140, 242
Precedence trees, 123
Preemption, 138
Prefix property, 120
Primality,
   co-RP, 310
   polynomial characterization, 298
Prim's algorithm, 135
   with Fibonacci heaps, 145
Principle of optimality, 56
Probabilistic algorithms: 29, 74,
   270
   complexity, 309–11
   isolation, 270
   maximal independent set, 293
   maximum matching, 28–30
   parallel random matching,
      270–74
   random connected components,
      85–88
   for vertex connectivity, 217
Problem transformation, 80–83
Processing element, systolic, 256
Processor utilization, 312
Pruning, 64, 228–29 (*See also*
   Backtracking)
$P(s)$, for synchronization, 275
Pseudopolynomial, 296, 309

Q

Queue operations, 154

R

Radius, 3
Random algorithms (*See*
   Probabilistic algorithms)
Random graphs:
   Erdos and Renyi, 86
   trees, 118–19
Random polynomial problems, 309
Reachability matrix, 10
Reachable from, 4, 11
Recurrence relations, 68–69
Recursion, 66–69, 158, 161, 232
Reducing path, 254
Register allocation, 141
   for trees, 143
Regular graph, 4–5, 17–19, 255
   planar, 53
Regular of degree r, 4, 17
Regular polyhedra, 235
Regular supergraph, 18
Representation of edges, 11–14
Representation (of graph), 8–14
   adjacency list, 11–14
   dynamic, 11–14
   linear array, 11–14
   linear list, 11–14
   pure linked, 11–13
   static, 8–10
Representation of vertices, 11–14
Resource allocation digraph, 139,
   244
R-factor, 239
R-factorable, 239
Ringleader selection, 311
RNC, 312
Root, in unary tree, 130
Rooted tree, 115
RP, 309

S

Satisfiability (SAT), 303
Scattering, in hypercube, 285
Search stack, 168
Search tree
   alternating, 21–26
   for backtrack coloring, 227–31
   backtracking, 63–64
   blocked, 21–23
   in Dijkstra's algorithm, 91
   in Ford's algorithm, 103
Sedgewick-Vitter heuristic shortest
   path algorithm, 107–9
Selection, tree processor, 279

117
fundamental minimal cuts
   representation, 116
Gallai's complementarity
   relations, 238
hamiltonian powers theorem, 46
Heawood map-coloring, 225
hungarian trees, 23
hypercube multiplication
   performance, 290
incidence transpose product, 10
isolating lemma, 270
Kuratowski's characterization of
   planarity, 33
linear bound on $|E|$ for planar
   graphs, 38
linear time heap creation, 129
linear time random components,
   87
logarithmic height unary tree, 131
logarithmic search times, 118
lower bound for parallel
   maximum, 241
LVP characterization, 311
matching algorithm performance,
   26
matching partition of bipartite
   graphs, 240
maximum flow equals minimum
   cut, 197
maximum flow/minimum
   capacity cut, 186
Mehlhorn error bounds, 72
minimax independent set of
   Turan, 240
minimum weight disjoint paths to
   minimum weight perfect
   matching, 83
monotonicity of Dinic's layered
   networks, 208
net flow and flow value, 185
noncontracting condition for
   spanning matching, 15
nonplanarity of Kuratowski
   graphs, 39
NP-completeness:
   complete subgraph, 305
   directed hamiltonicity, 307
   independent set, 306
   3-colorability, 306-7
   undirected hamiltonicity, 307
   vertex cover, 305
Ore's hamiltonicity, 45
orientability characterization, 163
parallel Dijkstra performance,
   262
parallel Floyd, 265
parallel heap creation, 278
parallel matching performance,
   274
partitioning regular bigraphs into
   matchings, 18
path and subpath optimality, 102

path diversity characterization of
   connectivity, 179
performance and optimality of
   Huffman encoding, 122
performance of Ford's, 106
performance under path
   compression and small-large
   rank rule, 134
Petersen's characterization of
   bridgeless cubic graphs, 239
Petersen's 2-factor condition, 239
pipe-organ optimality, 70
Redei's theorem, 46
regular graphs, 17
reliable transmission condition,
   182
scheduling lower bound, 244
search tree correctness, 22-23
spanning tree enumeration, 11
switch characterization of cross
   and back edges, 162
synchronization correctness and
   performance, for heaps, 277
time performance greedy II, 73
topological ordering, 137
trapping / non-trapping condition
   for orientability, 166
tree processor components, 282
TSP error bound, 75
Tutte matrix condition, 28
Tutte's hamiltonicity condition,
   48
Tutte's 1-factor condition, 239
upper bound, chromatic number,
   223
upper bound, edge chromatic
   number, 224
Valiant-Brebner hypercube
   routing, 220
validity of Dijkstra algorithm, 93
3-interchange, 74
3-optimization, 74
3-SAT, 304
Topological ordering, 137-38
Totally disconnected graph, 3
Trail, 3-4
Transitive closure, 4, 114, 244
   systolic, 259-62
Transitive reduction, 114
Trapping condition, 166
Traveling salesman problem, 45,
   51, 64, 69, 73-75
Tree, 4, 115-17, 148-49
Tree edge, 156, 159-60, 169
Tree processor, 279
Triangle inequality, 50
Turan, 240
Turing machine:
   accepting, 297
   deterministic, 296-97
   halt, 297
   language recognized, 297
   nondeterministic, 299

space complexity, 297, 299
time complexity, 297, 299
Turing table, 297
Tutte determinant, 28, 274
Tutte hamiltonicity condition, 48
Tutte hamiltonicity counterexample,
   48
Tutte matrix, 28
Tutte 1-factor condition, 239
2-optimization, 74
Type
   edge, 14, 16
   graph, 9-14, 16
   vertex, 14, 16

U

Unary representation, 274, 309
Unary tree, 130-32
Undirected graph, 3 (*See also*
   Graph)
Union (of graphs), 2
Uniquely n-colorable, 235

V

Vertex
   of a graph, 1,
   list representation, 8-14
Vertex chromatic number, 223
Vertex connectivity, 3, 178 (*See
   also* Connectivity)
Vertex cover, 237, 305
Vertex disconnecting set, 178
Vertex disjoint paths, 81, 83, 179
Vertex dominating set, 237
Vertex separator, 55
Visibility graph, 90
Vizing's inequality, 224, 308
Voronoi diagram, 76, 134, 148
Voronoi polygon, 76
$V(s)$, for synchronization, 275
Vulnerability, of a network, 181

W

Walk, 3-4, 9
Warshall's transitive closure
   algorithm, 114
Weighted digraph, 90
Weighted external path length, 118
Weighted graph, 125
Weighted internal path length, 118
Wigderson, approximate 3-coloring,
   236
Williamson planarity algorithm, 33

Z

Zero-knowledge passwords, 226-27